# THERAPY WITH CULTURED CELLS

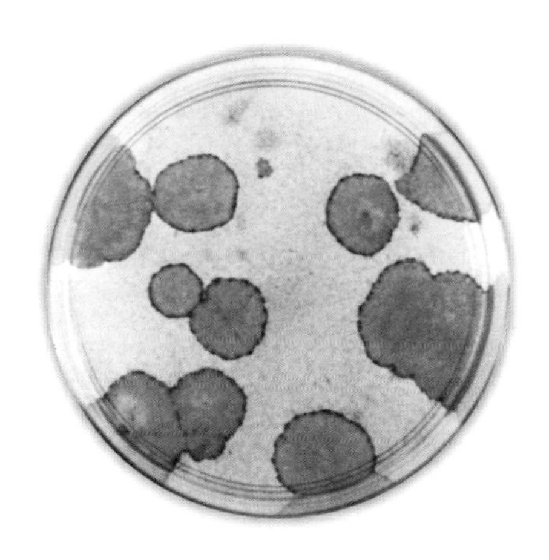

# THERAPY WITH CULTURED CELLS

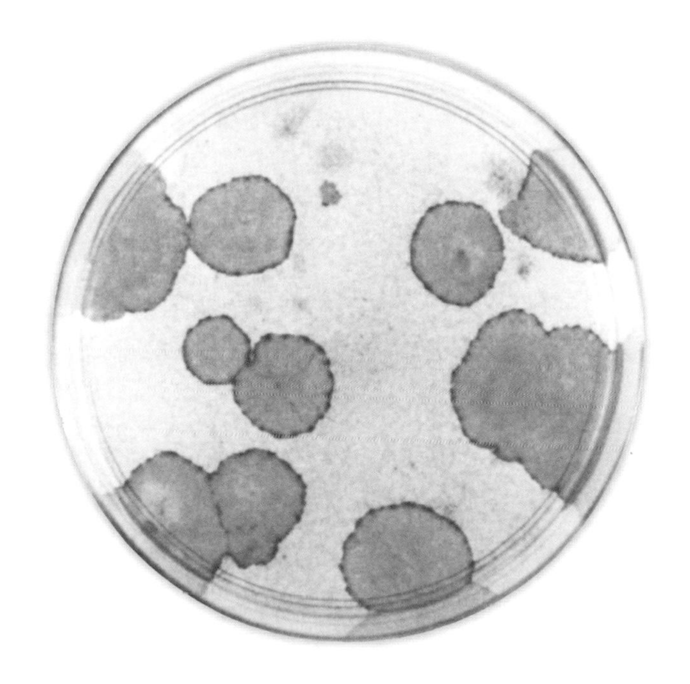

HOWARD GREEN
Harvard Medical School, USA

*Published by*

Pan Stanford Publishing Pte. Ltd.
Penthouse Level, Suntec Tower 3
8 Temasek Boulevard
Singapore 038988

Email: editorial@panstanford.com
Web: www.panstanford.com

**British Library Cataloguing-in-Publication Data**
A catalogue record for this book is available from the British Library.

**THERAPY WITH CULTURED CELLS**



ISBN-13 978-981-4267-70-0
ISBN-10 981-4267-70-8

Printed in Singapore by Mainland Press Pte Ltd.

# Contents

# Introduction

Ever since the birth of human therapy with cultured cells in my laboratory over thirty years ago, I have been occupied with the field of cell therapy. Since that time, the subject has attracted the interest and participation of scientists in the United States, France, Italy, Sweden, the Netherlands, Japan and Korea. It also attracted the participation of industry. In what follows, I will describe the early history of cell culture, the variety of cultured cell types used for therapy of different diseases, the large number of patients treated and the quality of the results obtained.

I will also include a description of embryonic stem cells and some of the problems that have to be solved before the somatic cells derived from them can be used for human therapy.

*Therapy with Cultured Cells* by H Green

www.panstanford.com
978-981-4267-70-0

Chapter One

# The Early History of Cell Culture

Since the early 19th century, the scientific world has been acquainted with the study of microorganisms in culture. But in order to study animal cells, it was necessary to isolate them from the organism and keep them alive *in vitro*. This became possible in the 20th century with the invention of tissue culture. Eventually this line of research led to discoveries in the fields of biology and medicine, in which culture of animal cells played an ever-increasing role.

## EXPLANT CULTURES

The first well-documented account of the cultivation of animal tissue *in vitro* is that of Harrison, at Johns Hopkins University. He demonstrated that embryonic tissue of the frog would develop normally in an explant culture (Harrison, 1907). In 1910, Alexis Carrel and Montrose T. Burrows, at the Rockefeller Institute, succeeded in cultivating explants of adult mammalian tissues (Carrel and Burrows, 1910a; 1910b).

## THE TRANSITION TO CELL CULTURE

Some years later, the first serial cultivation of dissociated mammalian cells was described (Rous and Jones, 1916). These

*Therapy with Cultured Cells* by H Green

www.panstanford.com
978-981-4267-70-0

investigators demonstrated that living tissue fragments could be dissociated with trypsin to produce cell suspensions that could be plated like bacteria, and that the cells could be subcultured at least twice.

## NUTRITIONAL NEEDS OF CULTURED CELLS

A long period of time elapsed before attention was first directed to defining the nutritional needs of cultured cells, so as to eliminate the use of complex media containing embryo extracts (Eagle, 1955). For this purpose Eagle used a cell line of mouse fibroblasts, Strain L (Sanford *et al.*, 1948) and a human uterine carcinoma cell, HeLa (Scherer *et al.*, 1953). Eagle was able to show that a total of 27 factors were necessary for growth, including 13 amino acids, glutamine, eight vitamins, six salts and a small amount of dialyzed serum protein. Under some circumstances, the requirement for serum protein could be satisfied with the addition of a polypeptide hormone (Hayashi and Sato, 1976). Other attempts to reduce the number of components were described by Ham and coworkers (Ham 1965; Hamilton and Ham, 1977).

Modern work on cultured cells is usually carried out with a medium that does not aim at minimal nutritional needs of the cells, but rather that is best suited for supporting the growth of a broad spectrum of mammalian cells. This formulation is based on the Dulbecco/Vogt modification of Eagle's Medium, which is provided commercially.

## USES OF CULTURED CELLS

1. For study of the action of viruses

   Such studies were carried out on Rous Sarcoma Virus (Temin and Rubin, 1958), SV40 virus (Todaro and Green, 1964), polyoma virus (Macpherson and Montagnier, 1964) and adenovirus (Freeman *et al.*, 1967).

2. For immunization to the poliomyelitis virus

   A key discovery that cell culture could be used to grow poliomyelitis virus was made by John Enders and his collaborators. They found that the virus could be grown in non-neural cells cultivated *in vitro*, such as those derived from human embryonic skin and muscle.

   They discovered that the virulence of the culture-grown poliomyelitis virus, when tested by cerebral inoculation in mice, declined markedly to a range between 100,000 and 1,000,000 fold (Enders *et al.*, 1949). These discoveries made possible a safe and effective method of immunization, which has remained in use up to the present time.

3. For study of cell movements

   Studies of migration of 3T3 cells were carried out by comparing their phagokinetic tracks with the orientation of actin- or tubulin-containing filaments. It was found that the tubulin-containing elements were oriented parallel to the direction of cell movement (Albrecht-Buehler, 1977).

4. For study of differentiated properties

   One example was the discovery of a preadipose murine cell line, 3T3-L1. These cells can be grown indefinitely in culture, but when their growth is arrested, they convert to adipose cells (Green and Kehinde, 1974; Green and Meuth, 1974) (Fig. 1). This cell line, which has been cited 79,000 times in the literature, and others like it (Green and Kehinde, 1976) are still used throughout the world to study various aspects of adipose differentiation.

5. For verification of GenBank data

   The gene encoding involucrin, a precursor of the cross-linked envelope of the terminally differentiated keratinocyte has, until recently, been undetectable in species outside the placental mammals. From GenBank data, it became possible to identify the involucrin gene in species as remote as marsupials and even birds.

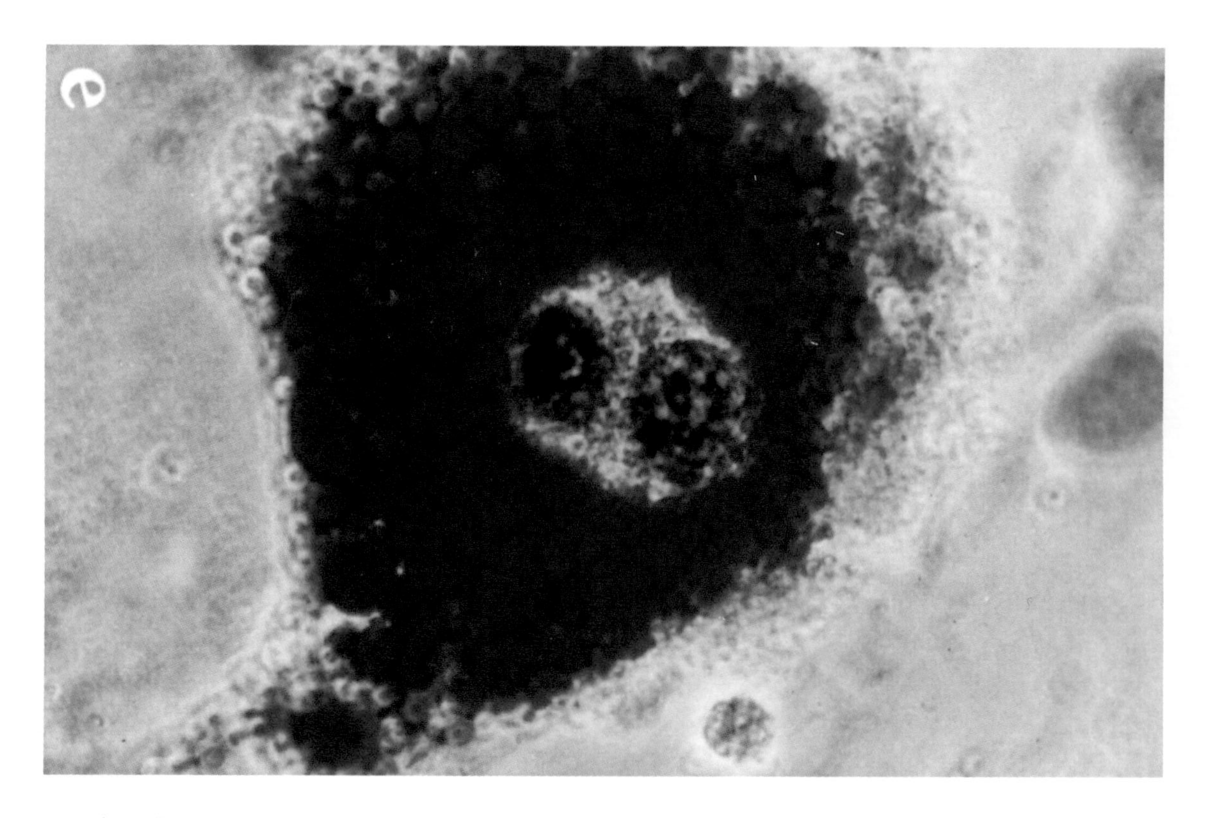


**Figure 1.** A single cell of line 3T3-L1 with massive accumulation of lipid stained with Oil Red O. The cell is binucleate, owing to failure of cytokinesis.

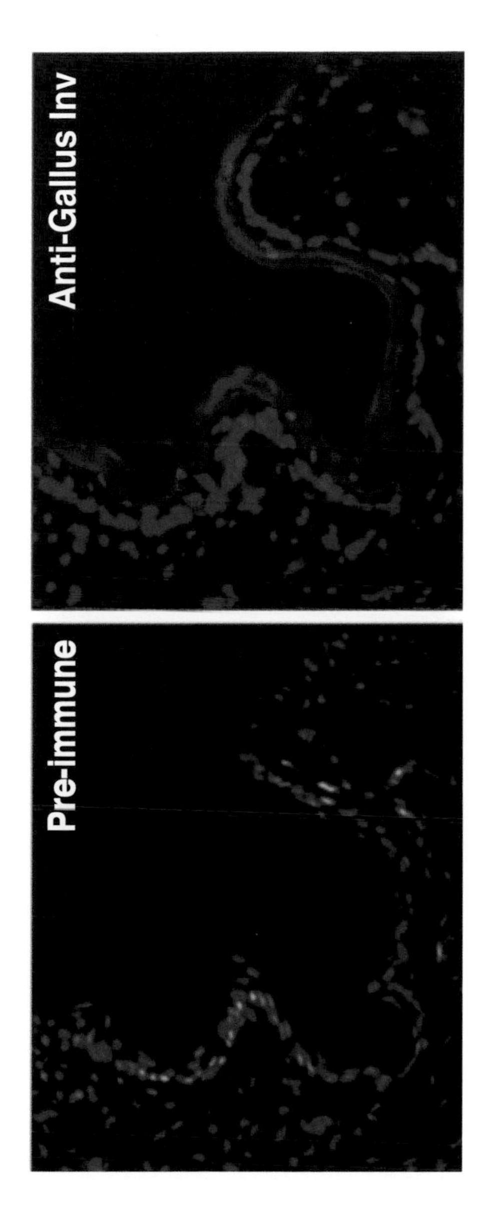


**Figure 2.** A section through Gallus skin was stained with involucrin (red) and counter-stained with DAPI (blue) for DNA. The cornified layer of the epidermis was strongly stained. The preimmuneserum gave only faint patchy background staining (Vanhoutteghem, Djian and Green, 2008).

But it seemed necessary to verify this conclusion experimentally. Sections were made through epidermis of the chicken (Gallus gallus) and stained with an antibody to the peptide sequence encoded by the Gallus involucrin gene. The result was that the cornified layer of the epidermis was strongly stained (Fig. 2)

6. For study of the chromosomal assignment of human genes

   Somatic cell hybrids containing human and mouse chromosomes tend to preferentially eliminate human chromosomes. This makes it possible to relate the expression of a human gene to the presence of a specific chromosome. In this way it was possible to make a chromosomal assignment of the human gene for thymidine kinase (Weiss and Green, 1967). Such studies also permitted an assignment of the poliovirus receptor gene to Chromosome 19 (Miller *et al.*, 1974).

   This kind of research was later rendered obsolete when the entire human genome was sequenced.

7. For production of monoclonal antibodies

   One of the most important discoveries made in this period was the development of monoclonal antibodies by fusing a mouse myeloma line with mouse spleen cells from an immunized donor (Kohler and Milstein, 1975). Such monoclonal antibodies have played an important role in medical treatment ever since.

8. For studies of disease due to the absence of keratinocyte transglutaminase

   Epidermal keratinocytes, late in their terminal differentiation, form cross-linked envelopes resistant to ionic detergent and reducing agent (Rice and Green 1977). The cross-linking process is catalyzed by the keratinocyte transglutaminase (TGase 1), which is anchored in the plasma membrane (Jeon *et al.*, 1998). Mutations in the TGase 1 gene result in the disabling congenital disease lamellar ichthyosis. The phenotype of this disease can be recognized by the inability of keratinocytes to form cross-linked envelopes in the patient's skin. Corneocytes scraped from the outer surface of the skin of a healthy individual possessed cross-linked envelopes that were insoluble in a solution of 2% sodium dodecyl sulfate and 2% 2-mercaptoethanol (Fig. 3A). Those scraped from the skin of a patient with lamellar ichthyosis, due to mutational absence of TGase 1, lacked insoluble envelopes (Fig. 3B). Therefore those cases of lamellar ichthyosis due to lack of TGase 1 may be readily distinguished from other ichthyotic diseases by a simple test for cross-linked envelopes (Jeon *et al.*, 1998).

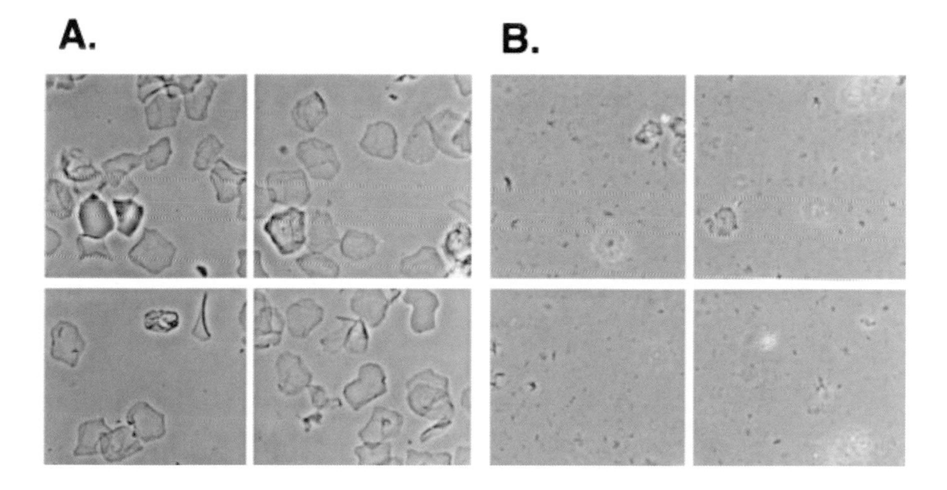


**Figure 3.** Absence of cornified envelopes in scales from a patient with TGase-negative lamellar ichthyosis. (A) Cornified cell envelopes in corneocytes scraped from normal human. (B) Absence of envelopes in corneocytes from a patient with lamellar ichthyosis.

9. For synthesis of therapeutic hormones

   Erythropoietin is a glycoprotein that is produced in the kidney and stimulates division and differentiation of erythroid progenitors in the bone marrow.

   In order to produce erythropoietin in cultured cells, the erythropoietin gene was introduced into CHO cells

(Chinese Hamster Ovary cells), which were then cultured in a serum-free medium. The cells synthesize and secrete the recombinant erythropoietin into the medium. The recovered erythropoietin is used for the treatment of anemia in humans.

10. For studies carried out in the laboratory of Elaine Fuchs on the stem cells of hair follicles and the demonstration of their self-renewal *in vitro* (Blanpain *et al.*, 2004).

11. Human diploid cells

    Cells obtained from mice, hamsters and other animals, when serially cultivated, become aneuploid and develop into established (immortalized) cell lines. In contrast to this behavior, fibroblasts of the human, when subjected to serial cultivation, do not undergo such changes. They may be grown through many population doublings but eventually lose all proliferative capacity (Hayflick and Moorhead, 1961). This stability of the human genome, which is characteristic of keratinocytes as well as fibroblasts, is what makes it possible to use keratinocytes for human therapy.

## Chapter Two

# The Beginnings of Keratinocyte Cultivation

As I have described (Green, 2008), in 1974 I never intended to work on the subject of cell therapy or the treatment of burns. At that time, current thinking was that much could be learned about embryogenesis from the study of cultivated murine teratomas, tumors that can give rise to all somatic cell types. In general, this hope was not fulfilled, but for me, studying cultures of serially transplantable teratoma derived by Leroy Stevens (Stevens, 1970) turned out to be a very fruitful enterprise. I was studying this cell line when I was distracted by an unexpected observation. From that point on, the direction of my research was guided by increasing familiarity with the material and what could be done with it. Perhaps the intellectual process that motivated me was similar to a description by Wernher von Braun, the German rocket physicist who, after the Second World War, came to the United States and played a leading role in the intercontinental ballistic missile program and the Apollo spacecraft that went to the moon. Von Braun said, "Basic research is what I am doing when I don't know what I am doing".

While studying cultures of Stevens' teratoma (Stevens, 1970), my student James Rheinwald and I discovered that the

*Therapy with Cultured Cells* by H Green

www.panstanford.com
978-981-4267-70-0

teratoma cells gave rise to colonies of epithelial appearance (Fig. 4) (Rheinwald and Green, 1975a).

But what kind of epithelial cells were they? When we isolated this cell type, made sections through its stratified colonies, and examined them by electron microscopy, we discovered that they belonged to the category of keratinocytes of stratified squamous epithelium. They contained desmosomes, keratohyaline granules and aggregated tonofilaments (Fig. 5).

We then wondered whether epidermal keratinocytes of the human, which had hitherto been uncultivable, would grow under the same conditions. When we performed the experiment, we were pleased to find that they grew very well indeed (Fig. 6) (Rheinwald and Green, 1975b).

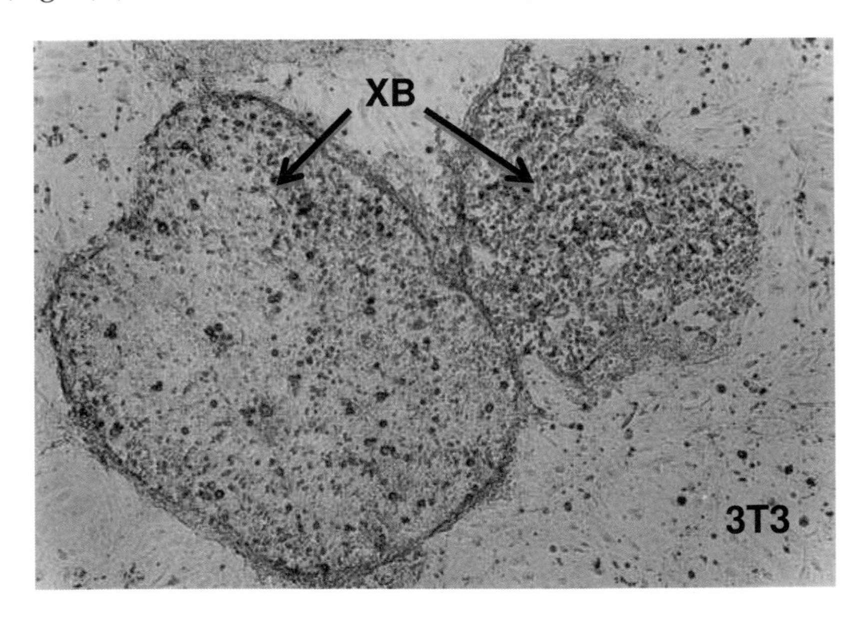


**Figure 4.** Clones of XB cells, an epithelial cell line derived from a murine teratoma, growing on a layer of lethally irradiated 3T3 cells.

What made both these discoveries possible was a study carried out many years earlier in my laboratory and which led to the development of the murine cell line 3T3. In 1961, a young medical student, George Todaro, came to work in my

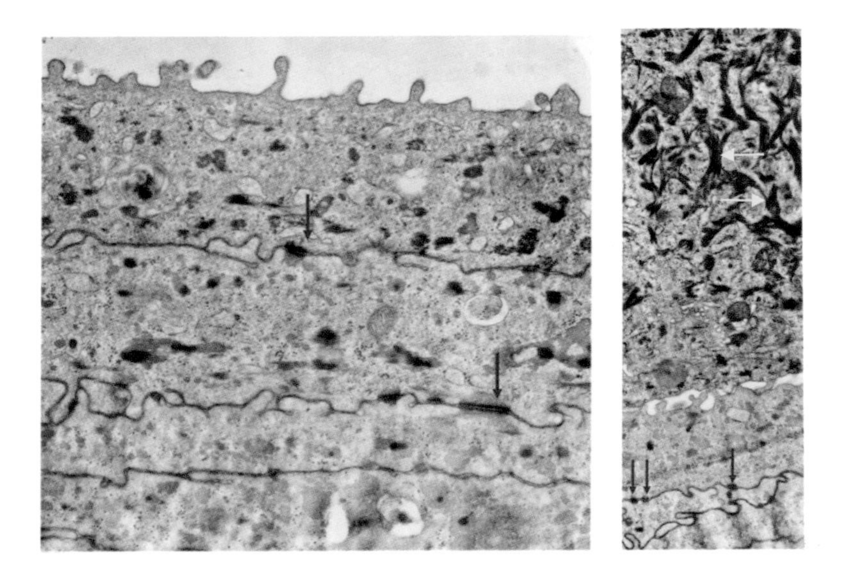

**Figure 5.** Desmosome indicated by red arrows, tonofilaments by yellow arrows.

laboratory at New York University Medical School. We began cultivation of mouse embryo fibroblasts with a view to establishing an immortalized cell line suitable as a target for viral transformation. At that time, it was believed that mammalian cells became immortalized in culture only rarely and that it was impossible to predict when such an event might occur or under what conditions. In order to avoid haphazard conditions of cultivation, I thought it necessary to keep both the inoculation density and the transfer interval constant during repeated subcultivation, because those two variables might influence the ability of the cells to become immortalized; in addition, knowledge of the correct conditions might make it possible to develop immortalized cell lines reproducibly.

We settled on inoculation densities of 3, 6 or 12 x $10^5$ cells per 20 $cm^2$ dish and a transfer interval of 3 or 6 days. After a period of declining growth rate that lasted for 10-20 cell generations, during which the doubling time of the murine fibroblasts increased to as much as 100 hours, we were pleasantly surprised to find that the growth rate began to increase

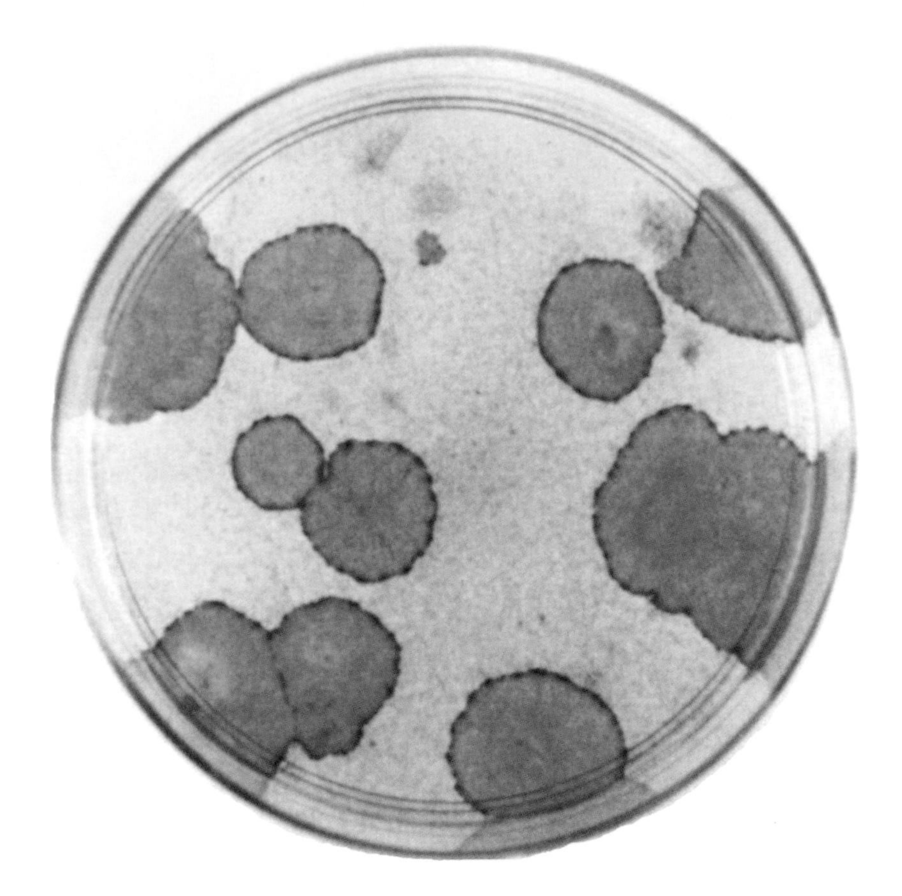

**Figure 6.** Colonies of human epidermal keratinocytes stained with rhodamine, against a background of supporting 3T3 cells.

and in 9 of 11 cultures carried under several conditions, there evolved cell lines with doubling times of 15-24 hours (Todaro and Green, 1963).

We were surprised a second time to discover that although immortalization occurred under most of the culture conditions we used, the properties of the resulting cell lines depended on the inoculation density and the transfer interval. The most interesting line resulted from serial subcultivation using the smallest inoculum. This line developed from cultures transferred every 3 days at an inoculation density of

$3 \times 10^5$ cells per 20 $cm^2$ dish, and accordingly was named 3T3. It grew as vigorously as the other cell lines as long as the cells were sparse, but arrested its growth sharply and entered a stable resting state when the cells became confluent at a saturation density of 50,000 cells/$cm^2$, only one sixth that of secondary cultures of strains of mouse fibroblasts. When transferred with dilution, the 3T3 cells resumed exponential growth, and again reached saturation density at 50,000 cells/$cm^2$. These experiments showed that murine fibroblasts maintained at low population density during the period of their immortalization evolved into a cell line which, when allowed to become confluent, entered a reversibly resting state at a very low population density.

In the earliest experiments by Rheinwald and myself on isolation of teratomal keratinocytes, cited above, we discovered that their growth in culture depended on the presence of fibroblast support and for this purpose we used 3T3 cells, lethally irradiated to prevent their own growth. This worked so well that it became the basis of all future work on human keratinocyte cultivation and the 3T3 cell line is used everywhere today to support the growth of cultured human keratinocytes. No other method produces results of comparable quality.

As soon as we learned how to grow human keratinocytes in culture (Rheinwald and Green, 1975b), we began to improve the culture conditions. The first agent we added to the culture medium was Epidermal Growth Factor (EGF), discovered by Stanley Cohen (Cohen and Elliott, 1963; Carpenter and Cohen, 1990). This polypeptide increased the culture lifetime of human epidermal cells from 50 to 150 cell generations (Fig. 7) (Rheinwald and Green, 1977).

Shortly afterwards, I began to examine the effects of cAMP on the proliferation of cultured keratinocytes. Hitherto it had been thought that cAMP brought about arrest of cell growth. But I found that dibutyryl cAMP promoted keratinocyte proliferation. The same was true for methyl isobutyl

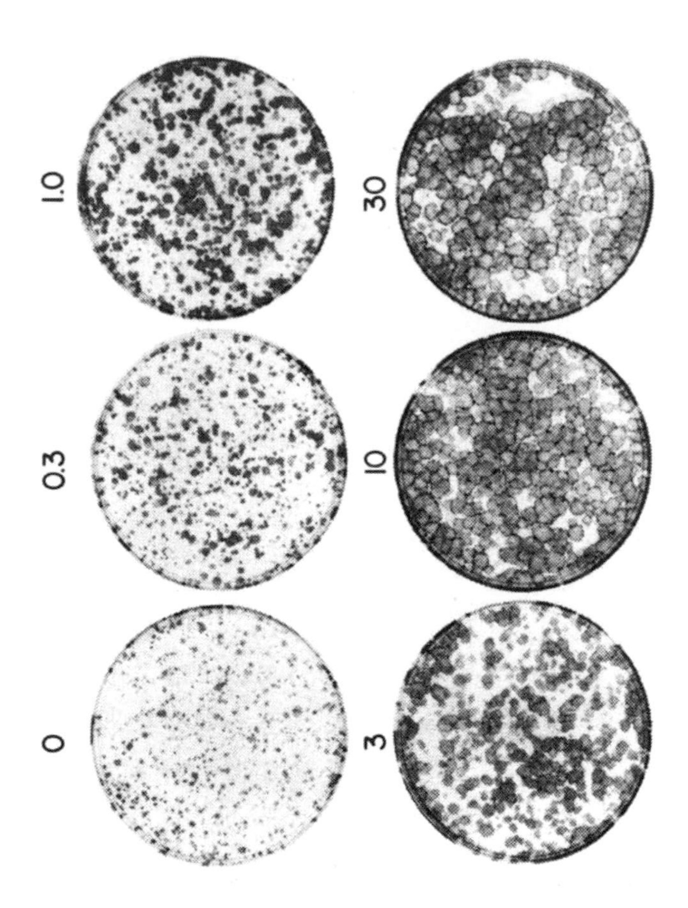


**Figure 7.** Effect of EGF on the size of keratinocyte colonies. EGF was added at indicated concentrations (in ng/ml) on day 4. Cultures were fixed on day 13. Colonies grown in the presence of EGF became progressively larger with increasing concentration up to 30 ng/ml (Rheinwald and Green, 1977).

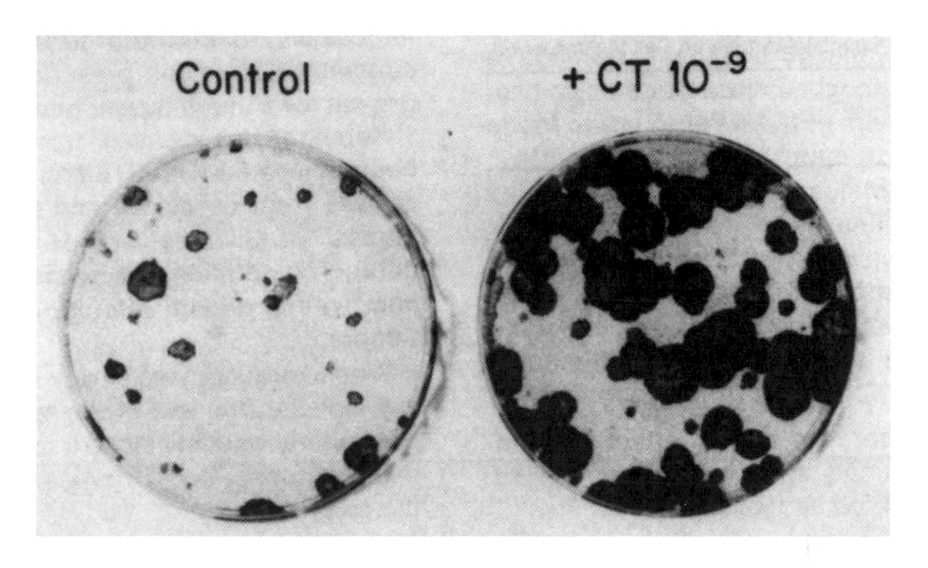


**Figure 8.** Effect of cholera toxin on colony growth.

xanthine, an inhibitor of phosphodiesterase and for isoproterenol, a well-known beta agonist. But the most effective growth-promoting agent was cholera toxin (choleragen) (Fig. 8) (Green, 1978).

Later, it was discovered by others that the addition of insulin, adenine, and Ham's nutrient mixture F-12 further improved proliferation (Allen-Hoffmann and Rheinwald, 1984).

Chapter Three

# The Treatment of Burns

As soon as we had optimized the culture conditions for the proliferation of human epidermal cells and realized that we could grow vast amounts from a tiny biopsy, I began to think of practical applications. Could we restore epidermis to burned patients by grafting them with cultures derived from a small biopsy of their own epidermis?

First we had to have a method of preparing grafts from cultures. The cells could not be simply scraped from the vessel surface because the basal layer, which contains the multiplying cells of the stratified culture, would be destroyed. After failing in several attempts to find a successful method of detachment, we hit upon the enzyme Dispase, a neutral protease discovered in Japan. This enzyme detached a confluent layer of cultured keratinocytes from the surface of the dish without dissociating the cells from each other. The detached cell layer then shrank to half or less of its area and could be picked up and, supported by a backing, be applied as a graft (Fig. 9) (Green *et al.*, 1979).

In order to demonstrate that such cultures could engraft, they were applied to full thickness wounds of athymic mice, which were incapable of mounting an immune response. The result was that the cultures engrafted successfully. This was evident by the formation of a characteristically thicker human epidermis and proven by the presence of human involucrin,

*Therapy with Cultured Cells* by H Green

www.panstanford.com
978-981-4267-70-0

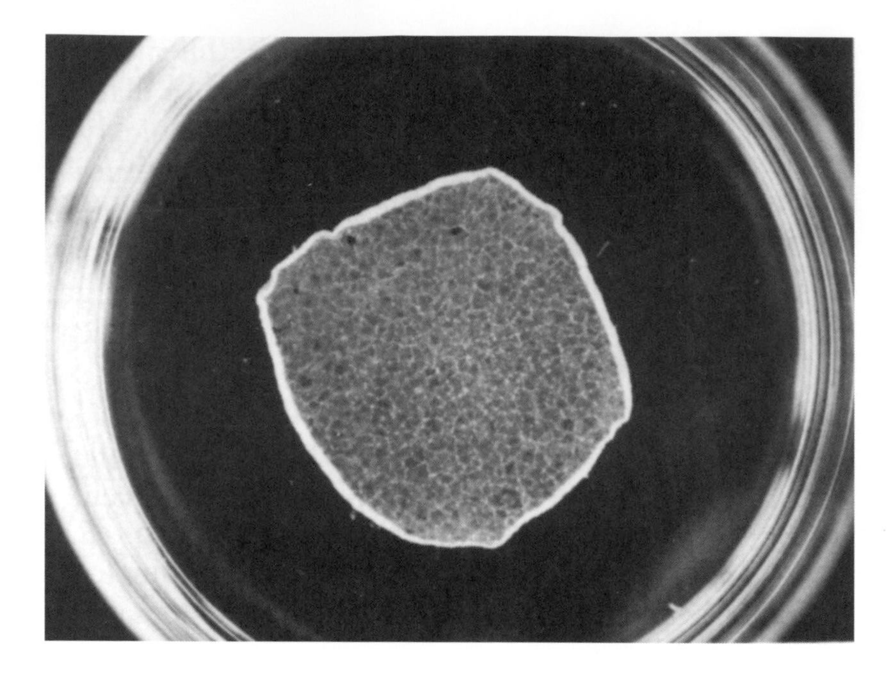

**Figure 9.** Cultured epithelium after its detachment from the dish with Dispase. The epithelium retains the curled edge of cells formerly on the wall of the dish.

which differs from that of the mouse and can be detected by antibodies specific for the human protein (Fig. 10) (Banks-Schlegel and Green, 1980).

I now felt that we could consider grafting cultures to humans suffering from 3rd degree (full thickness) burns. A colleague introduced me to Dr. Nicholas O'Connor, who was Director of the Burn Unit of the Peter Bent Brigham Hospital. I consulted with some friends and persons whom I knew by reputation, to discuss possible dangers. I received reassuring advice and decided to proceed.

Our first patient was a 61-year-old man whose arm had sustained 3rd degree burns. A small biopsy was taken from his unburned skin and 3T3-supported cultures were grown in my laboratory, at that time located at MIT. The cultures were transported by taxi to the hospital and applied. After a few days, it became obvious that some of the cultures

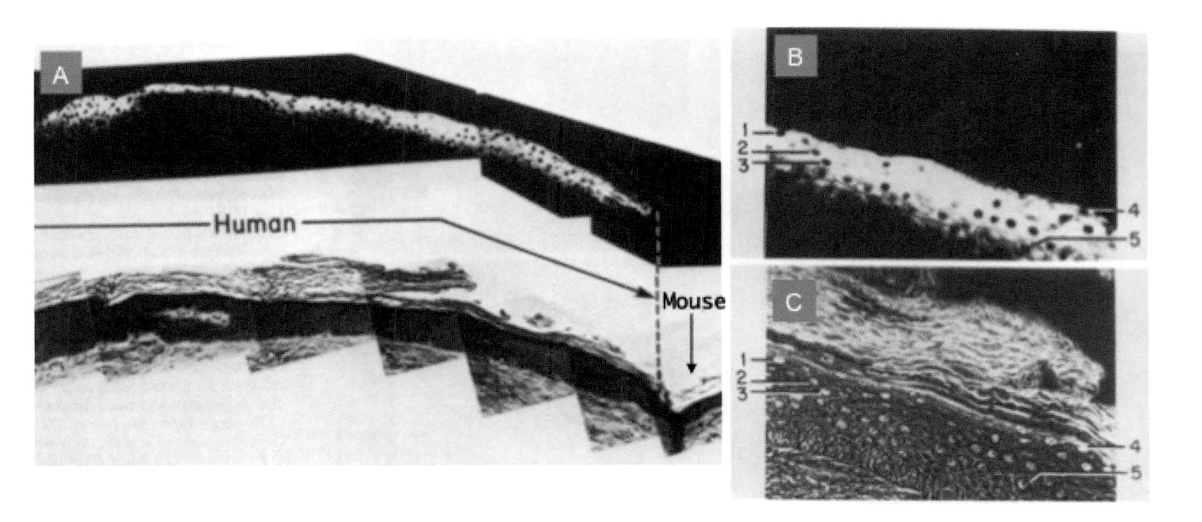


**Figure 10.** Human epidermis generated on an athymic mouse 108 days after application of a cultured human epidermal sheet. A. Note thickness of generated human epidermis compared with mouse epidermis. Only the human epidermis stains with a specific antibody to involucrin, a product of terminal differentiation of the keratinocytes. Comparison of B with C shows that involucrin is present in only the outer, terminally differentiating layers of the human epidermis generated from the culture. The thick stratum corneum, typical of the human but not of the mouse, is also shown in C. The numbers and arrows in B and C indicate identical positions in cell layers stained for involucrin [B] and those revealed by phase microscopy [C].

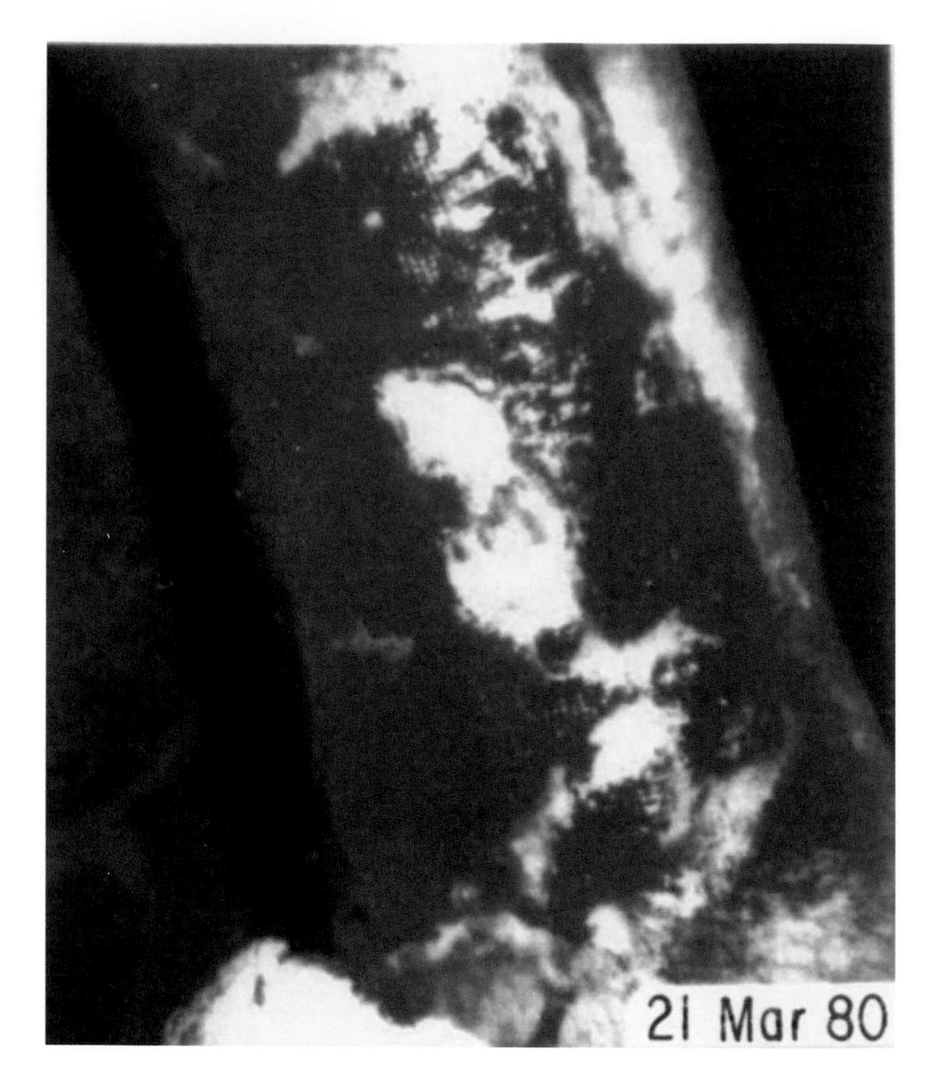


**Figure 11.** Regions of engraftment of Petri dish cultures of autologous epidermal cells to an excised surface of the first human treated. Some of the engrafted regions preserve the circularity of the Petri dish cultures.

had engrafted (Fig. 11). Repeated applications of the cultures resulted in complete coverage of the burned area (O'Connor *et al.*, 1981).

We continued to grow cultures for occasional burn cases without much change in the importance of the procedure until 1983, when two brothers, 5 and 6 years of age, living in Wyoming, were burned over 97 and 98% of their body surface, mostly of third degree. Dr. John Remensynder, Director of the Shriner's Institute of Boston told me that the brothers had no chance of survival with conventional treatment, but if I would try to save them with cultured cells, he would accept them in transfer from Wyoming. I was not prepared for the scale of the necessary cultivation but of course, I had to try and both brothers were repeatedly grafted with cultures prepared in my laboratory, which by then had moved to Harvard Medical School (Fig. 12). The two brothers survived and lived for over 20 years after returning to their home in Wyoming (Gallico *et al.*, 1984). They would have lived longer but for medical complications not directly related to the burns or the grafting. The experience with the two brothers demonstrated clearly that the grafting of cultures was life-saving.

The use of cultured autologous cells for the treatment of burns and scars was investigated very early in Japan. It was shown that under favorable conditions the cultures adhered well and generated skin of normal appearance (Kumagai *et al.*, 1988).

Other applications for cultured autologous epidermal cells were soon discovered. Giant congenital nevi are a premalignant condition and must be excised. In small children, these nevi commonly cover one third or more of body surface. Cultured epidermal autografts are valuable for the restoration of the excised surface. In 16 operations on eight patients the cultured epithelium remained as permanent robust skin (Gallico *et al.*, 1989).

In congenital hypospadias the penile urethra terminates on the ventral surface of the penis. Treatment requires reconstruction of the penile urethra. Squamous epithelial cells derived from the urethral meatus can be cultivated and used to restore the anterior urethra (Romagnoli *et al.*, 1990; Romagnoli *et al.*, 1993).

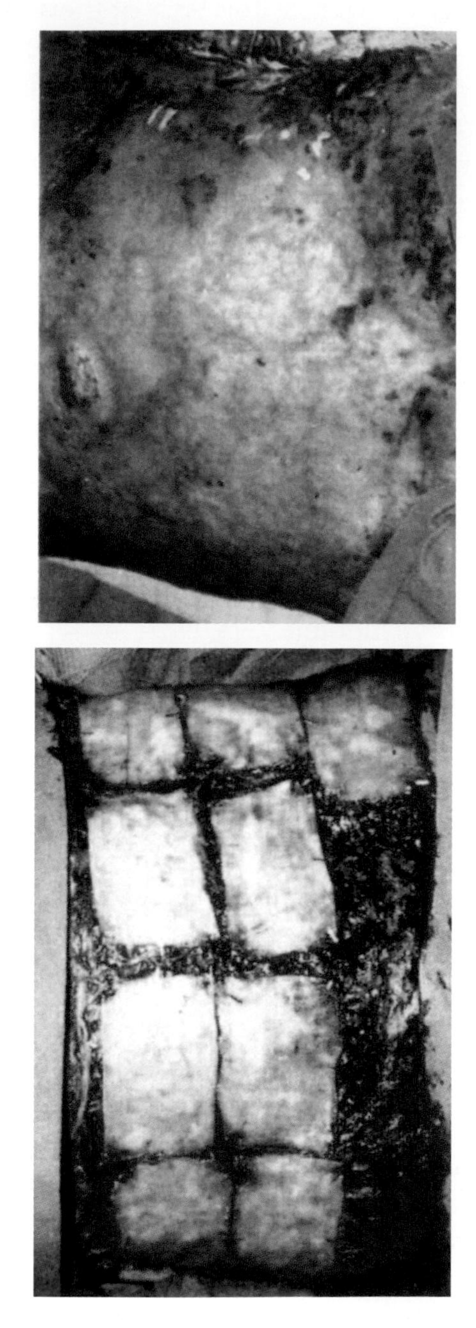

**Figure 12.** *Left*, abdomen of one of the brothers treated by application of now rectangular grafts to his abdominal surface after excision to muscle fascia. *Right*, 14 weeks later, when the surface was covered with thick confluent epidermis (Gallico *et al.*, 1984).

Melanocytes are difficult to grow in pure culture, but they grow well in the presence of keratinocytes. Patients with stable vitiligo or piebaldism have been treated successfully with cultures containing autologous melanocytes growing in association with keratinocytes. When the epidermis affected by stable vitiligo is excised and cultures are applied to the created partial thickness wound, the healed epidermis becomes populated by melanocytes and these persist for at least 7 years (Guerra *et al.*, 2000; Guerra *et al.*, 2003; Guerra *et al.*, 2004).

Achromic regions of the knees and legs, and of the ankles and feet were removed by a pulsed Erbium:YAG laser. Grafts of keratinocytes and melanocytes prepared from cultures one day after reaching confluence were applied immediately after removal of the epithelium (Guerra *et al.*, 2000). The results were evaluated 6 months later, when there was 76-90% restoration of pigment.

An example of complete recovery is shown in Figure 13.

## THE REGENERATION OF DERMIS

At this time, some surgeons believed that a healthy epidermis depended on the dermis and if the dermis were destroyed by a 3rd degree burn, grafting of epidermal cells alone would not generate a robust epidermis. It was necessary, they said, to graft a composite culture containing some sort of dermal substitute together with the epidermal cells (Green, 1989). Of course the dermis has a very complex structure and it is impossible to imagine how such a structure could be fabricated. This did not deter attempts to make feeble imitations, some of which even became commercial.

This idea was soon proven to be entirely wrong. A study of the skin regenerated from epidermal cultures alone showed that the regeneration of the dermis occurred spontaneously with time (Compton *et al.*, 1989). A fully stratified epithelium developed very soon after grafting. Rete ridges appeared at the dermo-epidermal junction of the regenerated skin and

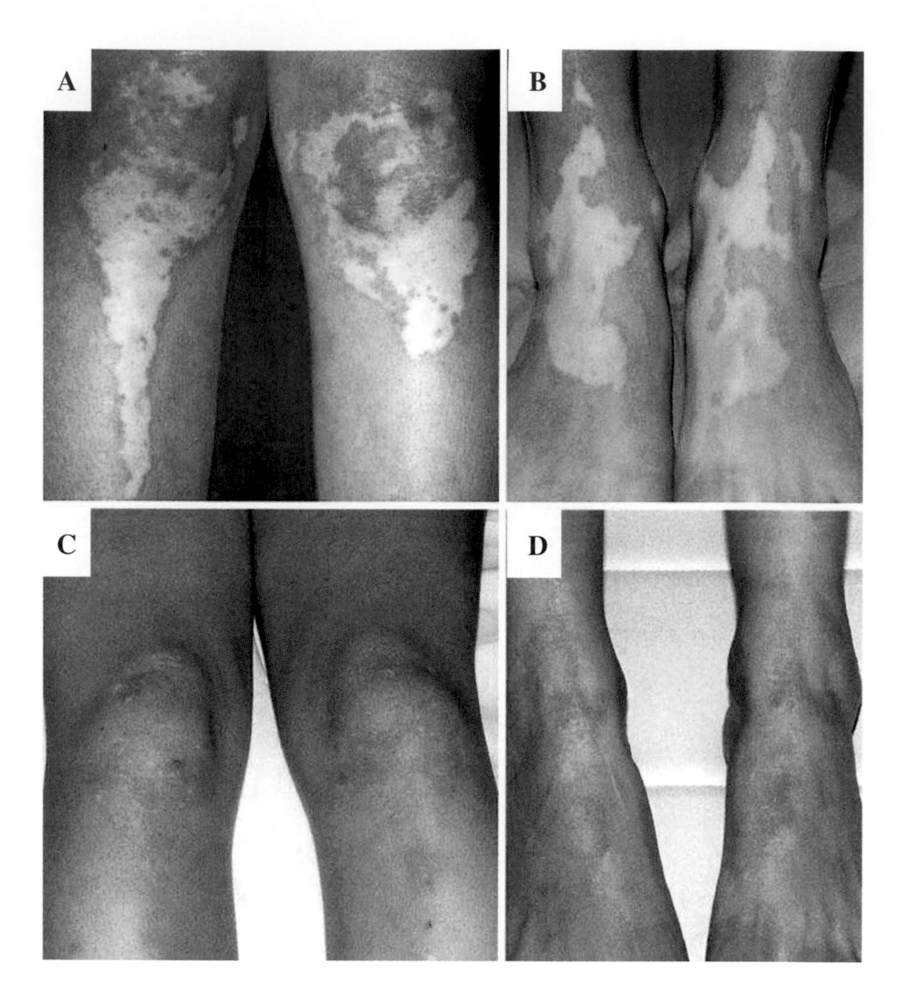


**Figure 13.** (A) achromic regions of knee and legs. (B) achromic regions of the ankles and feet. (C) complete repigmentation of the knees and legs. (D) Complete repigmentation of the ankles and feet (Guerra *et al.*, 2003).

became progressively more normal over a period of years. The sub-epidermal connective tissue was remodeled to produce papillary and reticular dermis, with fine collagen fibers in the sub-epidermal region and thicker fibers below. After one or two years, the number and size of the anchoring fibrils resembled those of normal skin. All the features of the dermis continued to improve with time, until by five years after grafting, the dermis appeared completely regenerated (Fig. 14).

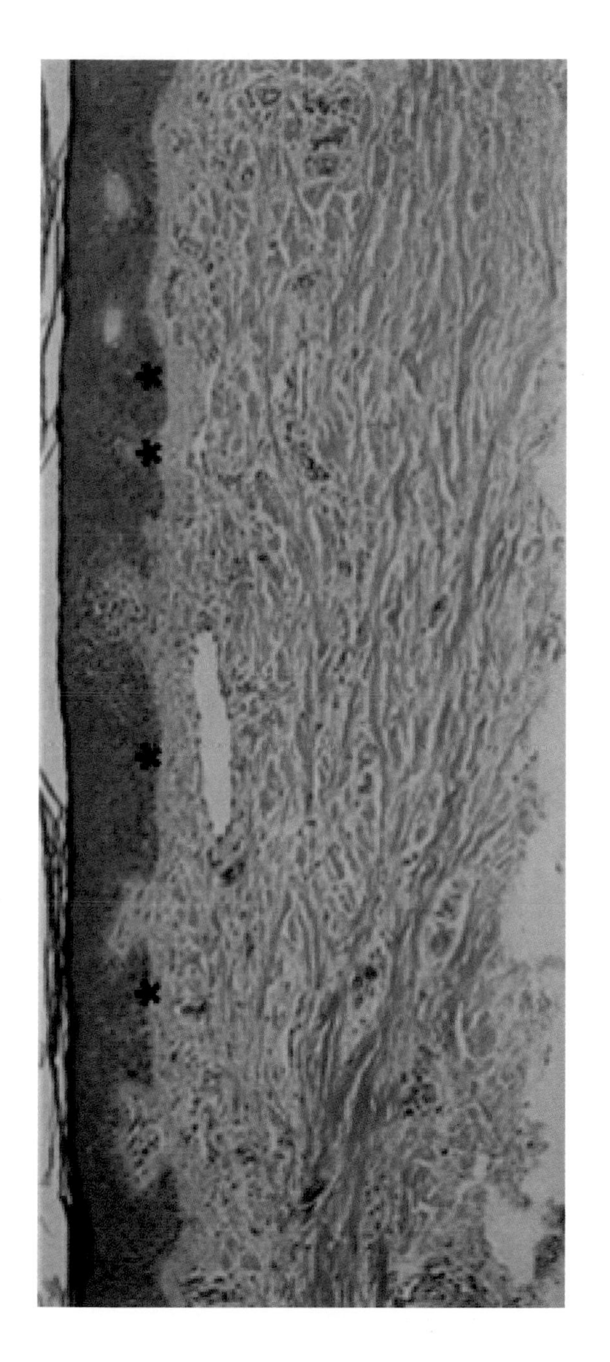

**Figure 14.** The regenerated epidermis appears normal, with regularly spaced rete ridges (stars). The subepithelial connective tissue is bilayered, with finer collagen bundles superficially and coarser collagen bundles beneath, together with normal vascular architecture (from Compton *et al.*, 1989).

## THE ROLE OF INDUSTRY

After the case of the two burned brothers from Wyoming, described above, it became obvious that I could no longer grow in my own laboratory the necessary cultures for burn patients. In 1987 a company called BioSurface Technology was formed in Cambridge, MA to provide cultures for burn victims. This company was later absorbed by Genzyme, Inc., which continues to this day to provide autologous cultures for treatment. Over a period of five years, Genzyme sent autologous cultures to the French Military Burn Hospital at Clamart, near Paris, where the cultures were applied to patients. As a result of their study at the Burn Hospital, it was concluded that the cultures provided by Genzyme were highly beneficial and permitted a higher survival rate of the patients. Genzyme has treated over 1500 patients since 1994. It now treats 90 new patients per year. Most of the patients are in the United States, but quite a few are in France and some are in Greece.

In 2008, I met a woman who had been treated for burns many years earlier. In 1995, she was in her office in Georgia when a small airplane crashed just outside and skidded into her building, killing everyone on board and sending flaming airplane fuel into her office. She was burned over 86% of her body surface. Epidermis was restored over the burned area by multiple grafts of autologous epidermis prepared by Genzyme. I was pleased to meet the lady and to find that she still had a pretty good life thirteen years later (Fig. 15).

## A PHILOSOPHICAL REFLECTION

Some kinds of research, if not carried out today by someone, will be carried out by someone else next year.

What I have described here is not research of that kind. It depended on a unique coincidence of time and place and on my willingness to follow a path open at that time.

If I had not followed that path, the Birth of Therapy with Cultured Cells might not have taken place then, or ever, given the regulatory inventions that followed.

For this reason, I feel grateful that I had that opportunity and that the results, 34 years after the research began, are what they are today.

**Figure 15.** Shirley Badke.

At the end of 2002, the company Tego Science was formed by my former post-doctoral fellow, Saewha Jeon in Seoul, Korea for the treatment of burns, nevi and scar revision. Between 2003 and 2008, Tego Science treated 134 patients suffering from burns with cultured autologous keratinocytes. They applied 6,323 grafts, with a take rate of 63 to 89%. Since then Tego Science has treated a small number of patients for scar removal, vitiligo, aplasia cutis congenita and ulcers.

In 2007, the Japan Tissue Engineering Company (J-TEC) in Aichi, Japan, received regulatory approval for cultured epidermal grafts for the treatment of burns (Fig. 16). In Japan, there are 300 cases per year of full thickness burns affecting over 30% of body surface. At present, J-TEC is treating one new patient per week.

## FIBRIN AS A SUPPORT FOR CULTURES

This subject has a long history. The use of fibrin glue as a substrate for cultured autologous cells was first described as a faster procedure for preparing grafts in the treatment of two burn patients (Ronfard *et al.*, 1991). A study of thirty patients was later carried out with cells grown on a fibrin substrate, with favorable results (Carsin *et al.*, 2000). This method was studied extensively in the laboratory of Yann Barrandon at the Ecole Normale Supérieure in Paris (Ronfard *et al.*, 2000). First they showed that the fibrin matrix did not reduce the growth potential of the keratinocytes. They then applied the method to seven patients with extensive burns. They concluded that the method had numerous advantages over earlier methods. The time required to produce graftable cultures was shortened. The cultures were easier to prepare and easier to use by the surgeon.

Other advantages of growing keratinocytes on a fibrin matrix have been described (Pellegrini *et al.*, 1999; De Luca *et al.*, 2006).

1. A keratinocyte culture on a plastic dish shrinks to half or less of its area after detachment (Fig. 16) and has to be attached to a gauze backing for shipping and handling by the surgeon. A culture grown on a fibrin matrix does not shrink and requires no backing for handling (Fig. 25).
2. Keratinocytes cultured on a fibrin matrix have the same growth capacity and stem cell content as those cultured on plastic surfaces, but the enzymatic detachment is avoided.
3. The use of a fibrin matrix permits a reduction of the minimum time between biopsy and graft generation from 21 to 17 days. This is because cultures on a fibrin matrix need not be confluent for handling and application to wounds.

(a)

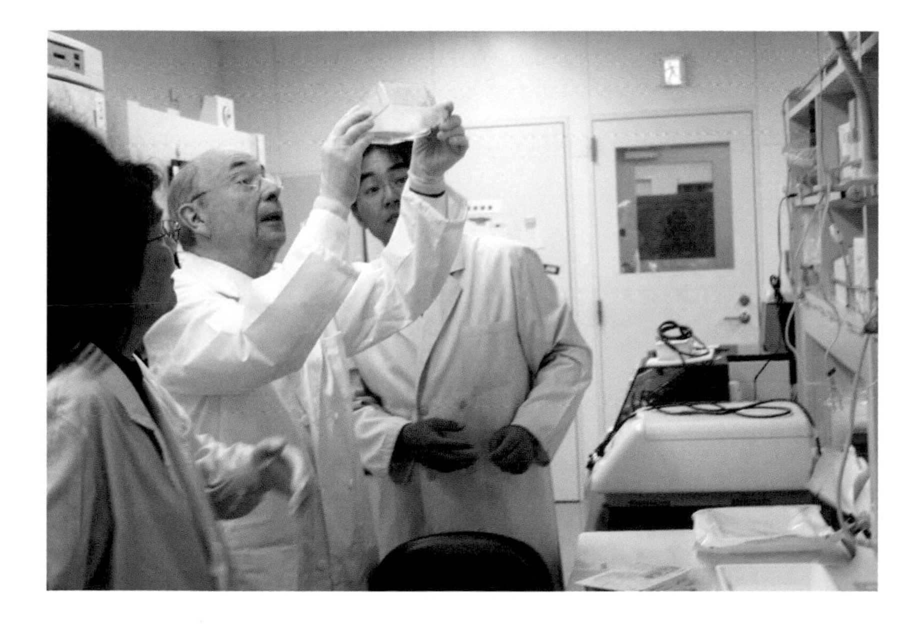

(b)

**Figure 16.** (a) New building of J-TEC. (b) I inspect cultures at J-TEC.

4. A keratinocyte culture grown to confluence on a plastic dish does not retain clonogenic ability for more than a short time after its detachment, limiting the time available for long distance transportation before grafting. In contrast to enzymatically detached cultures, which are usually transported at 4°C to prevent terminal differentiation, fibrin supported cultures can be safely transported at room temperature, as they are not confluent.

**Chapter Four**

# Defining the Stem Cell Character of Keratinocytes

Some time after therapy with cultured keratinocytes was demonstrated, Y. Barrandon and I carried out a detailed study of the growth potential of individual keratinocytes. We discovered that colony-forming human epidermal cells are heterogeneous in their capacity for sustained growth. Once a clone was derived from a single cell, its growth potential could be estimated from the colony types resulting from a single plating, and the clone could be assigned to one of three classes. The holoclone has the greatest reproductive capacity: under standard conditions, fewer than 5% of the colonies formed by the cells of a holoclone abort and terminally differentiate (Fig. 17). The paraclone contains exclusively cells with a short replicative lifespan (not more than 15 cell generations), after which they uniformly abort and terminally differentiate. The third type of clone, the meroclone, contains a mixture of cells of different growth potential and is a transitional stage between the holoclone and the paraclone. The incidence of the different clonal types is affected by aging, since cells originating from the epidermis of older donors give rise to a lower proportion of holoclones and a higher proportion of paraclones.

*Therapy with Cultured Cells* by H Green

www.panstanford.com
978-981-4267-70-0

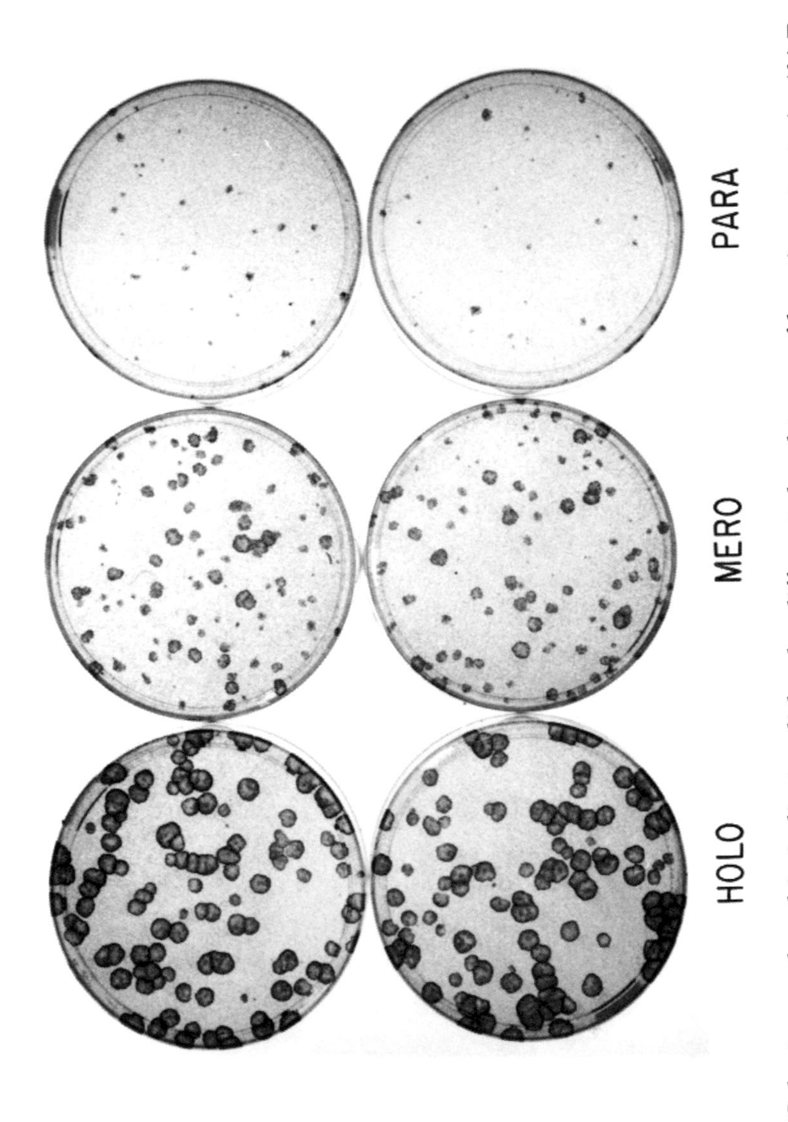


**Figure 17.** Colonies produced in indicator dishes by different clonal types of keratinocyte strain AY. Each clone was disaggregated, and one-quarter of the cells was inoculated into each of two indicator dishes containing irradiated 3T3 cells. The cells were fixed and stained on Day 12. Note the reproducibility of the colonies in duplicate platings.

## THE ROLE OF P63 AS A STEM CELL DETERMINANT

The protein p63 is a transcription factor with remarkable properties. Ablation of the gene encoding this protein in mice results in one of the most interesting phenotypes to be found in mammalian development (Yang *et al.*, 1998; Mills *et al.*, 1999; Yang *et al.*, 1999; McKeon, 2004). The affected offspring are born with essentially no stratified squamous epithelia of any subtype. Although several related epithelial cell types, such as mammary, urethral, prostatic, urothelial and tracheobronchial are also affected, the action of the protein is restricted to a rather narrow spectrum of epithelial cell types. In the human, even heterozygous mutations in the gene result in epithelial abnormalities, particularly in the cranio-facial region (Celli *et al.*, 1999; van Bokhoven and McKeon, 2002).

The nature of the function of p63 in development of squamous epithelia is still the subject of controversy, but there is substantial evidence supporting the hypothesis that p63 is a stem cell determinant in these epithelial cell types. The basis for this hypothesis, elaborated in the laboratory of Frank McKeon, is that in the newborn mouse whose p63 gene has been ablated, there are detectable terminally differentiated suprabasal keratinocytes, but the proliferative basal layer containing the stem cell population necessary to sustain the epithelium is lacking (Yang *et al.*, 1999; Senoo *et al.*, 2007).

The p63 gene generates transactivating N-terminally truncated transcripts (ΔNp63) initiated by different promoters. Alternative splicing give rise to three different C-termini, designated $\alpha$, $\beta$ and $\gamma$ (Yang *et al.*, 1998) (Fig. 18). In human epidermis and epidermal cultures, the p63 protein was found to be restricted to cells with high proliferative potential, located in the basal layer (Parsa *et al.*, 1999). The 4A4 monoclonal antibody detects all splice forms of p63 (Yang *et al.*, 1998), but it has become clear recently that one isoform, designated $\alpha$, the most abundant isotype in the mouse embryo (Yang *et al.*, 1999), gives the most precise identification of the stem cells (Di

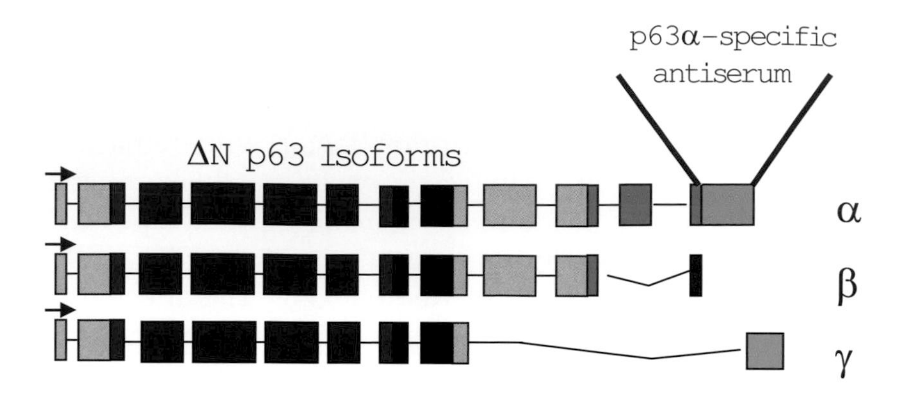


**Figure 18.** Structure of the ΔN isoforms of p63.

Iorio *et al.*, 2005). The abundance of the $\alpha$ isoform — but not that of the $\beta$ and $\gamma$ isoforms — strongly declines during the clonal transition from holoclone to meroclone and the protein is virtually absent from paraclones.

The identification of holoclones by immuno-detection of p63, especially the $\alpha$ isoform, is an important and simple means of determining the presence of an adequate number of stem cells in a cultured keratinocyte graft. This criterion has been applied to the use of cultured grafts of ocular limbus (Di Iorio *et al.*, 2005).

Chapter Five

# Treatment with Allogeneic Cultured Keratinocytes

Studies of the wound-healing effects of cultured allogeneic keratinocytes began many years ago (Hefton *et al.*, 1983; Hefton *et al.*, 1986; Thivolet *et al.*, 1986).

During the following years, many investigators studied the use of allogeneic keratinocytes in healing partial thickness burns, ulcers and other wounds. They found that the cultures had impressive therapeutic value (Leigh *et al.*, 1987; Phillips and Gilchrest, 1989; Phillips *et al.*, 1989; Teepe *et al.*, 1990; Beele *et al.*, 1991; Phillips *et al.*, 1991; De Luca *et al.*, 1992; Marcusson *et al.*, 1992; Teepe *et al.*, 1993a; Teepe *et al.*, 1993b; Duinslaeger *et al.*, 1997; Harvima *et al.*, 1999; Khachemoune *et al.*, 2002).

At first there was some confusion as to how the wound-healing effects of allogeneic keratinocytes were exerted (Shehade *et al.*, 1989). Some investigators believed that no immune response could be mounted against the allogeneic cells and that they were incorporated into the healed region. This was soon shown to be incorrect as study of healed chronic ulcers failed to reveal any DNA derived from the allogeneic cells (Brain *et al.*, 1989; Burt *et al.*, 1989; Phillips *et al.* 1990; Roseeuw *et al.*, 1990). It became clear from this work that the therapeutic effect of the allogeneic keratinocytes was due to their transitory presence in the healing wound.

*Therapy with Cultured Cells* by H Green

www.panstanford.com
978-981-4267-70-0

During the period 1996-2000, a series of articles published by Walid Kuri-Harcuch, Professor of Cell Biology at the Center of Investigation and Advanced Studies (CINVESTAV), together with his collaborators in Mexico City, shed much light on the usefulness and mechanism of the therapeutic action of allogeneic keratinocytes. They showed that skin donor sites and deep partial thickness burns underwent more rapid healing when banked allografts were applied (Rivas-Torres *et al.*, 1996). In a controlled study of deep partial thickness burns, they showed that frozen banked allogeneic cultures not only accelerated healing, but also made it possible to heal wounds that would otherwise not heal at all (Alvarez-Diaz *et al.*, 2000).

One example of treatment of a deep partial thickness burn is shown in Figure 19. Nine days after treatment with allogeneic keratinocytes, the wound was healed. The control, treated with conventional dressing, remained unhealed.

Allogeneic keratinocytes promote healing when applied after facial dermabrasion for removal of acne scars (Fig. 20).

They also studied the usefulness of allogeneic keratinocytes in healing chronic leg ulcers in ten patients. These ulcers were as large as 100 $cm^2$ in area and of up to twenty years duration. Some ulcers were so deep as to expose tendons. All of these ulcers were healed (Fig. 21). (Bolivar-Flores and Kuri-Harcuch, 1999).

This group developed an animal assay to demonstrate the therapeutic effect of human allogeneic keratinocytes. They made full-thickness wounds in the skin of immunocompetent mice. The untreated wounds reepithelialized from the wound edge at a rate of 150 $\mu$m/day. When cultured epidermal sheets were applied to the surface of the wound, the murine epithelium advanced at a rate of 267 $\mu$m per day. Most wounds treated with cultures healed after 10 days whereas untreated wounds required 16 days (Tamariz *et al.*, 1999).

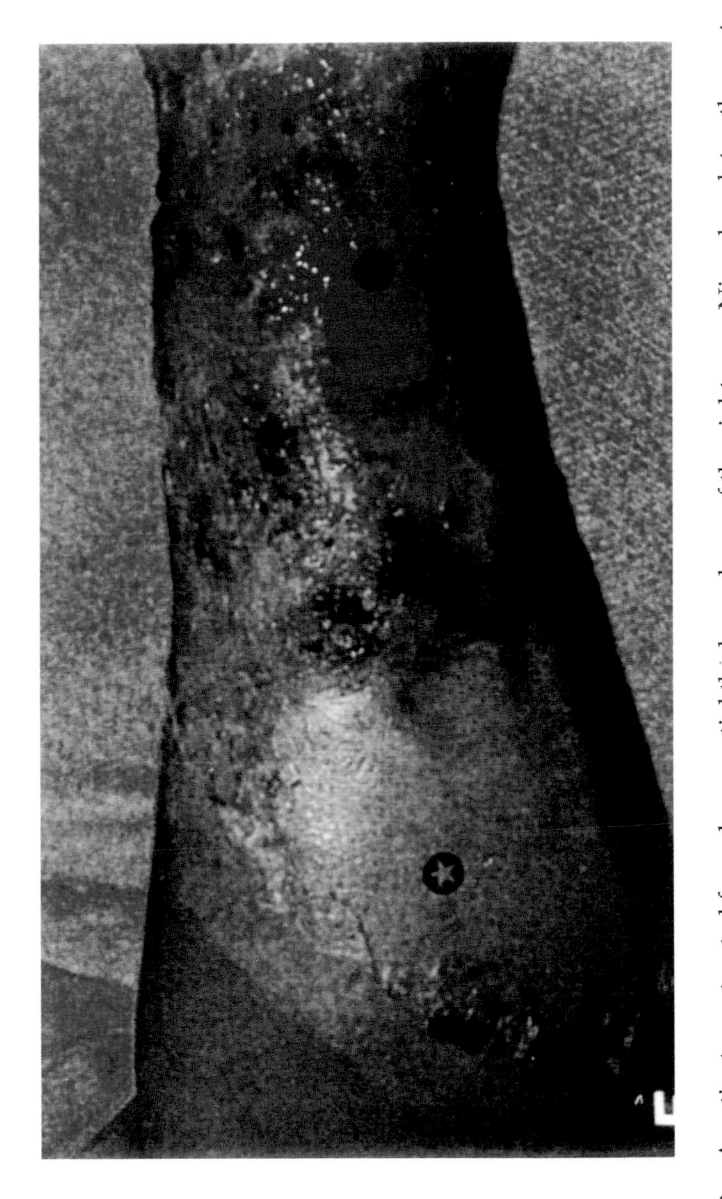

**Figure 19.** A patient was treated for a deep partial thickness burn of the right arm. Nine days later, the part of the arm treated with allogeneic keratinocytes had healed (left). The adjoining part (right), treated with conventional dressing, remained unhealed (courtesy of Professor W. Kuri-Harcuch).

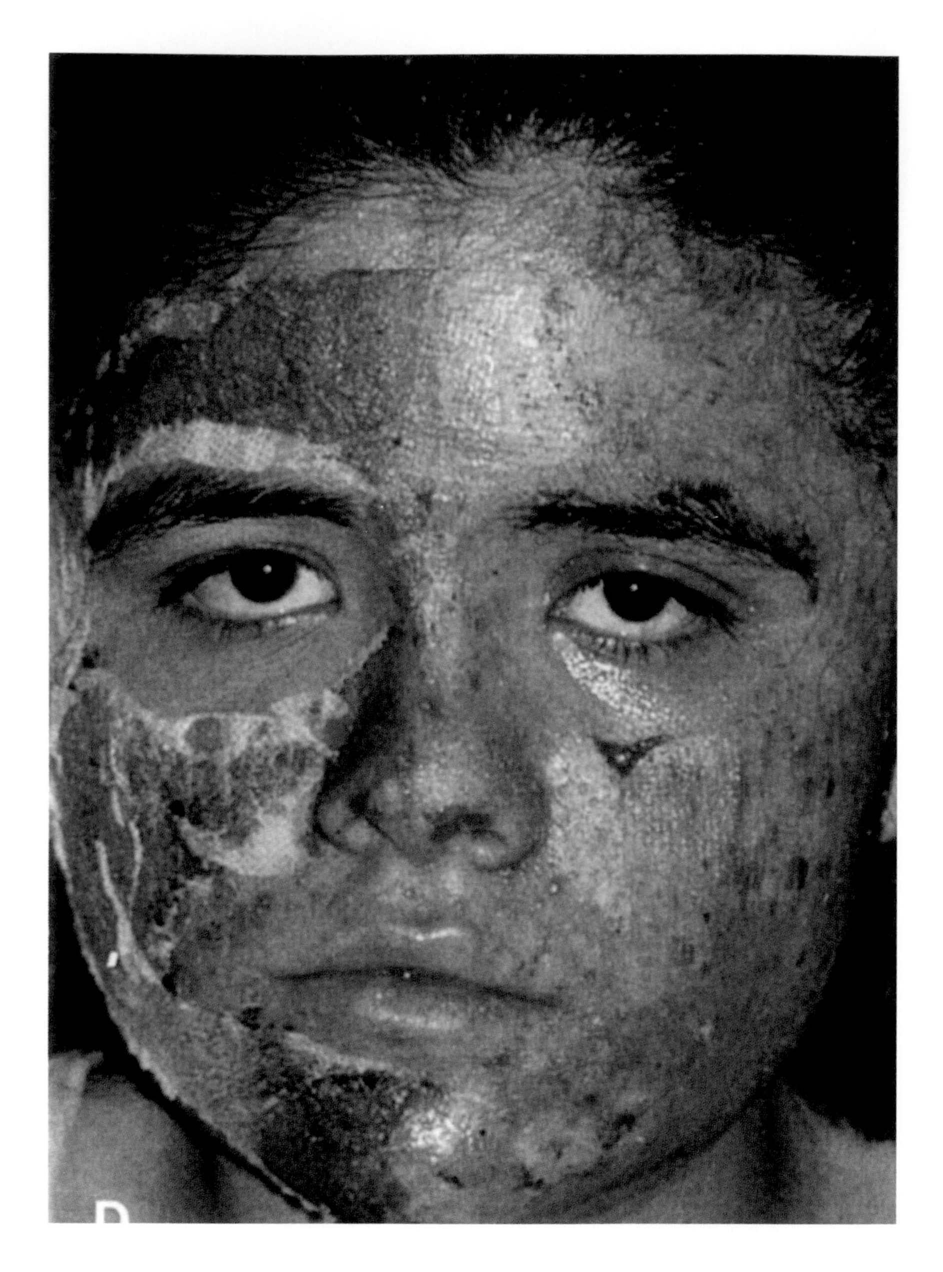

**Figure 20.** A patient three days after dermabrasion. The left side of her face, treated with allogeneic keratinocytes was completely healed. The right side, treated with conventional dressing, remained unhealed (courtesy of Professor W. Kuri-Harcuch).

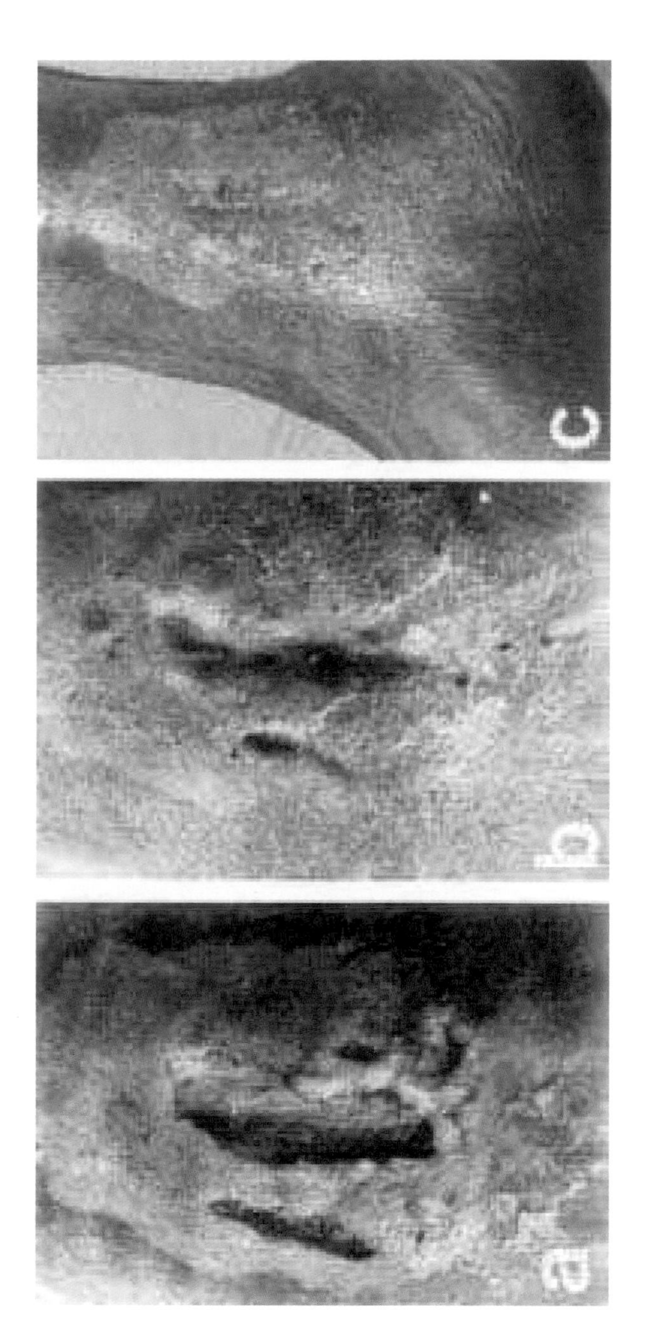


**Figure 21.** Patient 3. Chronic leg ulcer of 20-years' duration. (A) Before debridement. (B) After the third application of a thawed culture (5 weeks); epithelium has advanced from the edges of the wound toward its center. (C) Epithelization and complete healing of the ulcer after the fourth application of thawed culture (7 weeks).

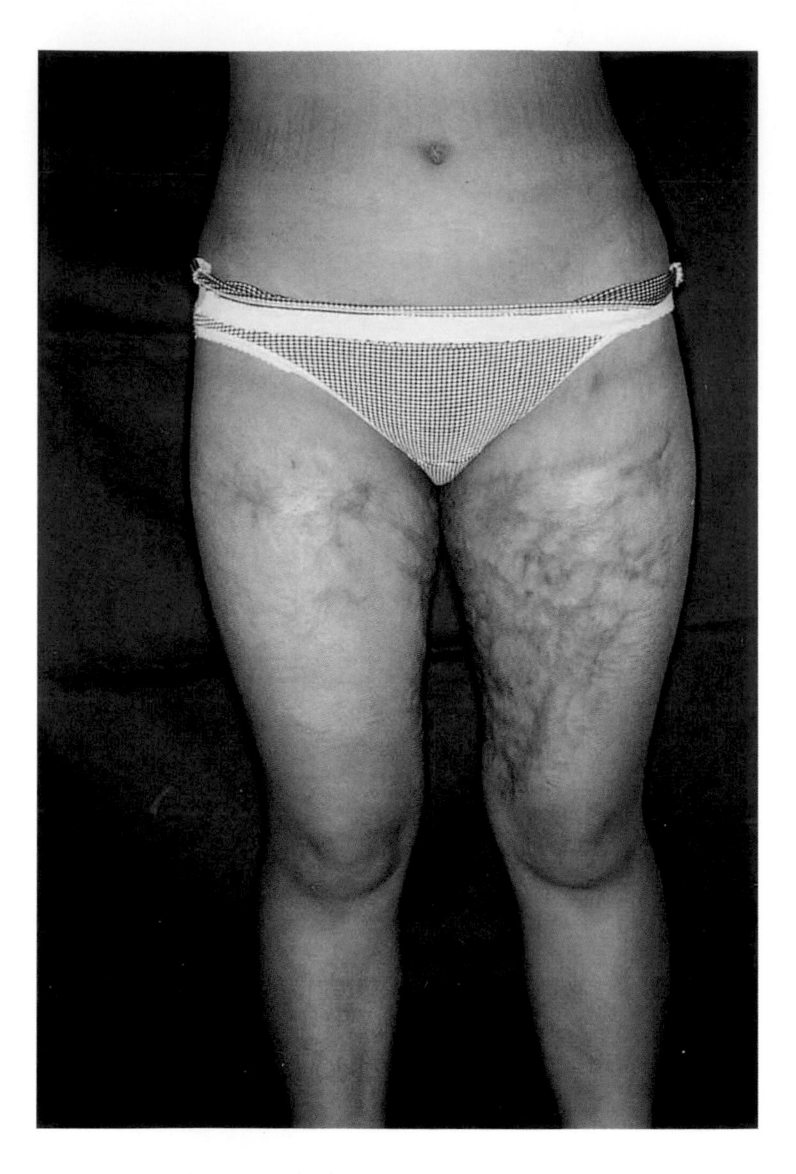

**Figure 22.** An eight-year-old female fell into boiling water and sustained scald burns. The right thigh was treated with cultured allografts; the left thigh was treated only by debridement. Six years after grafting, the right thigh has milder scar than the left thigh, which received only debridement (Yanaga, 2001).

Later it was shown that the application of cryopreserved cultured epidermal allografts to deep partial thickness burns and to split-thickness skin donor sites not only accelerated healing, but also suppressed scar formation (Fig. 22). (Yanaga *et al.*, 2001).

The question soon presented itself, what properties of the allogeneic cells were required in order to exert their therapeutic effect? Since banked keratinocytes frozen at −20°C contain no cells capable of proliferation, the acceleration of wound healing by such cultures does not require retention of proliferative ability of the keratinocytes (Bolivar-Flores and Kuri-Harcuch, 1999; Tamariz *et al.*, 1999). To this day the mechanism of the therapeutic effect of allogeneic keratinocytes remains unclear.

In spite of the facts that the effectiveness of therapy with allogeneic cultures has been conclusively demonstrated and that their use is without risk, allogeneic cultures have not been made available by any commercial/industrial entity up to the present day. This is a sad reflection, given all the contemporary emphasis on the importance of research on human therapy.

## Chapter Six

# Treatment of Ocular Disease

When scientific discovery opens up a new field of research it is impossible to anticipate where it may lead. I now wish to turn to some unanticipated consequences of the discoveries that I have just described.

We knew from early experiments on cultivating keratinocytes of all stratified squamous epithelia (esophageal, oral, vaginal etc.) that corneal keratinocytes grew poorly on sub-cultivation. The reason for this was elucidated in an extremely important article from the laboratory of T.T. Sun (Schermer *et al.*, 1986). From an analysis of the keratins of corneal cells and the surrounding limbal cells, as well as other data, the authors postulated that the stem cells of the cornea were located in the limbus, which surrounds the cornea and separates it from the conjunctiva (Fig. 23).

Chemical burns of the eye, if they destroy the limbal stem cells, can lead to a very nasty condition of inflammation, pain and loss of vision. The limbus can be restored by grafts from the healthy eye to the affected one (Kenyon and Tseng 1989). This requires large transfers of limbal tissue from the healthy eye. But limbal stem cells belong to the same family as epidermal keratinocytes and can therefore be cultivated in the same way, starting with a tiny biopsy. This was first shown in the laboratory of J. Rheinwald (Fig. 24). (Lindberg *et al.*, 1993).

*Therapy with Cultured Cells* by H Green

www.panstanford.com
978-981-4267-70-0

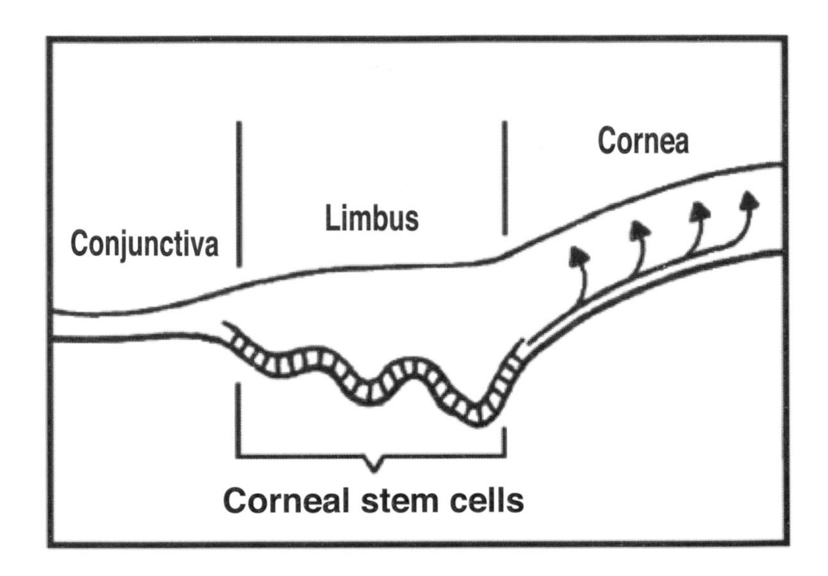


**Figure 23.** From Schermer, Galvin and Sun, 1986.

A few years later it was demonstrated in the laboratory of M. De Luca and G. Pellegrini, at that time in Genoa, that limbal cultures could be used therapeutically. In two patients suffering from alkali burns of a single eye they took a 1-2 mm biopsy from the limbal region of the healthy eye, grew the cells in culture, prepared a graft and applied it to the suitably prepared injured eye (Pellegrini *et al.*, 1997). In both patients, the results were the formation of a clear corneal epithelium, no neovascularization and great increase in visual acuity, over a follow-up period of two years.

## THE USE OF FIBRIN FOR OCULAR THERAPY

Later, Pellegrini and De Luca introduced the use of a fibrin support for the limbal cultures (Fig. 25). In collaboration with numerous ophthalmologists (Rama *et al.*, 2001), they obtained relief of symptoms in 14 of 18 patients. With subsequent repair of deeper injury by penetrating keratoplasty, vision could be totally restored (Fig. 26). Within five years, 116 patients had been treated. Grafting of autologous limbal cells was

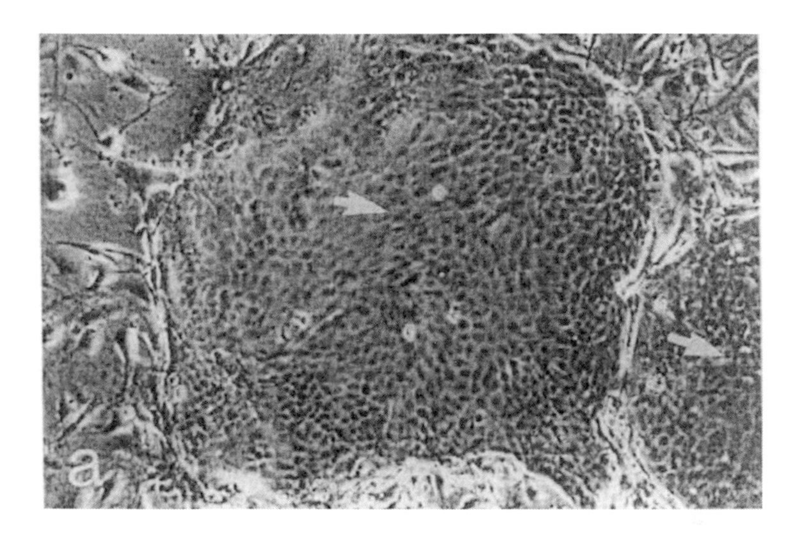


**Figure 24.** Photomicrograph of epithelial cells cultured from limbus. The epithelial cell colonies (white arrow heads) are surrounded by 3T3 supporting cells, which are displaced from the dish as the epithelial colonies expand (Lindberg *et al.*, 1993).

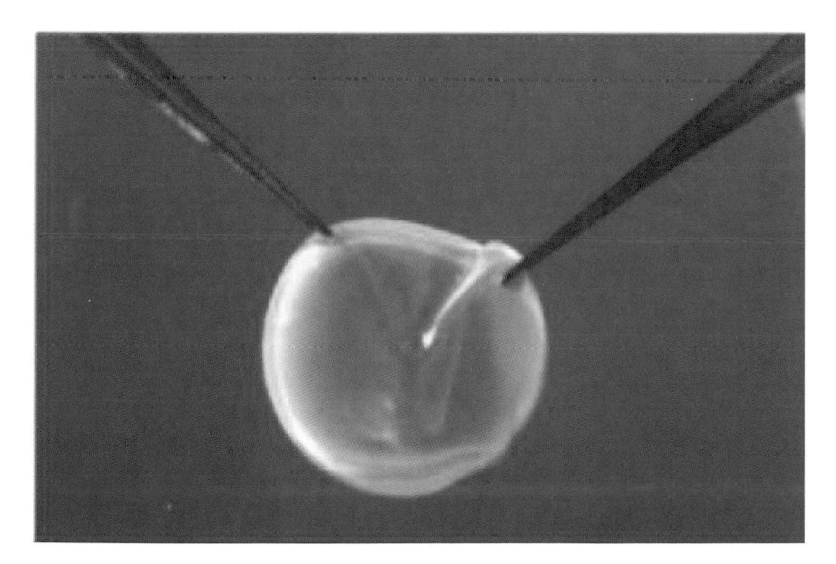

**Figure 25.** Secondary culture of limbal cells on a fibrin matrix of approximately 3 cm diameter. The fibrin is completely digested within 24 h of grafting, leaving no intervening matrix between the limbal cells and the wound bed (De Luca *et al.*, 2006).

successful in 78% of patients. Continuing integrity of the regenerated corneal epithelium was confirmed by up to three years of follow-up (De Luca *et al.*, 2006) and more recently up to nine years of follow-up (Pellegrini *et al.*, 2009).

The use of cultured autologous limbal cells was extended to patients with destruction of limbal/corneal epithelium due to abuse of contact lenses and ensuing bacterial infection. In five out of six patients, the procedure was highly effective in relieving symptoms and restoring vision (De Luca *et al.*, 2006).

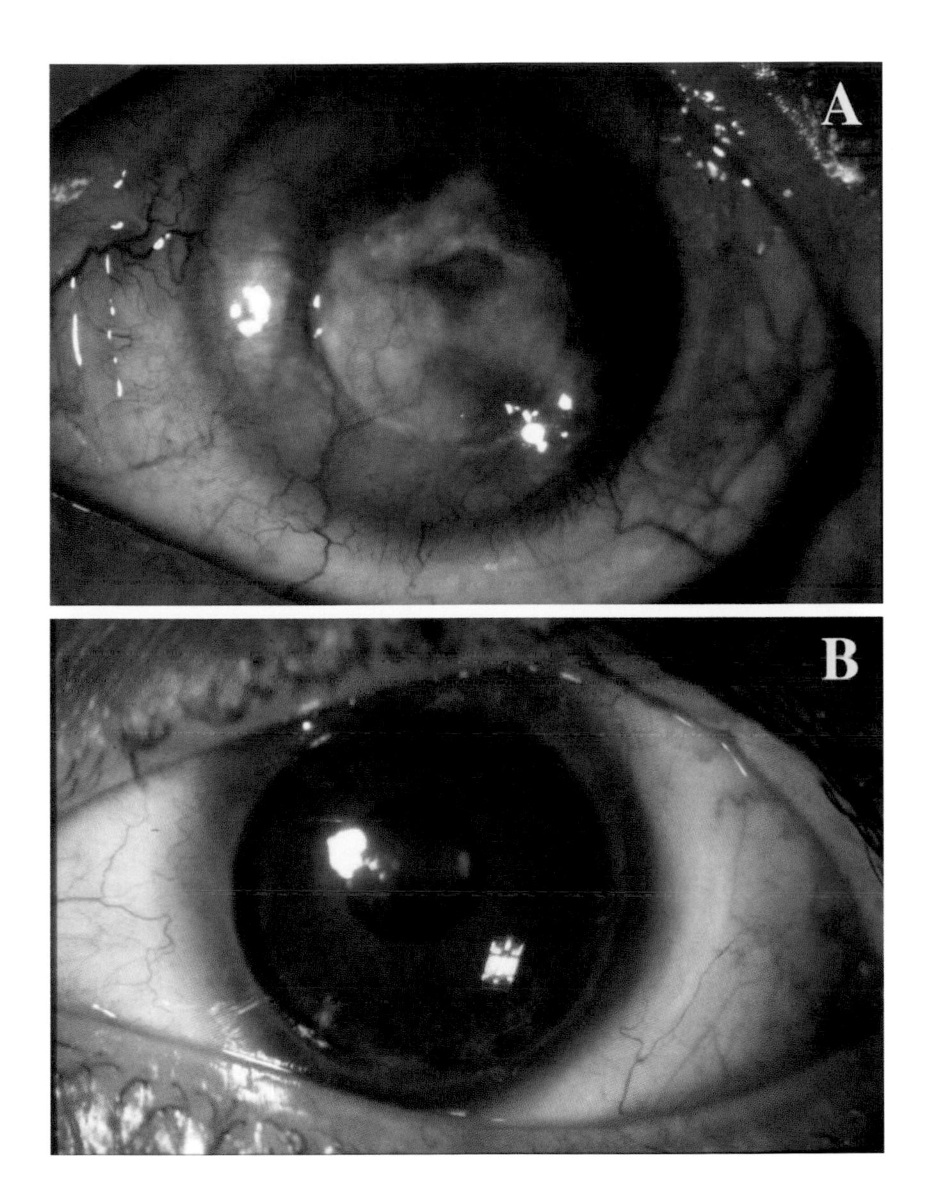


**Figure 26.** (A) Before treatment. (B) Five years after graft of limbal culture and four years after corneal transplant to remove stromal scarring (Rama and Pellegrini, 2001; De Luca *et al.*, 2006).

## Chapter Seven

# Gene Therapy

This discovery was another contribution of the laboratory of M. De Luca and G. Pellegrini (Mavilio *et al.*, 2006). They studied a severe genetically determined blistering disease of the skin — Junctional Epidermolysis Bullosa. The patient they studied was a double heterozygote containing a frame shift /single point mutation in the gene for Laminin 5-$\beta$3, which links basal epidermal cells to the basement membrane.

They cultivated epidermal cells of the patient and transduced them with a retroviral vector bearing a full length Lam5-$\beta$3 cDNA under control of the Moloney virus LTR. They then removed some affected skin and replaced it with grafts of transduced cells (all treatment with virus was ex vivo). As shown in Figure 27, the disease was cured. In principle, this form of therapy is applicable to other genetically determined blistering diseases.

This form of therapy has several advantages. First, patients with Epidermolysis Bullosa commonly get skin cancer in the affected region. This risk is eliminated by the treatment. Second, if any undesirable complication were to arise with the treated area, the transduced cells could be easily removed.

*Therapy with Cultured Cells* by H Green

www.panstanford.com
978-981-4267-70-0

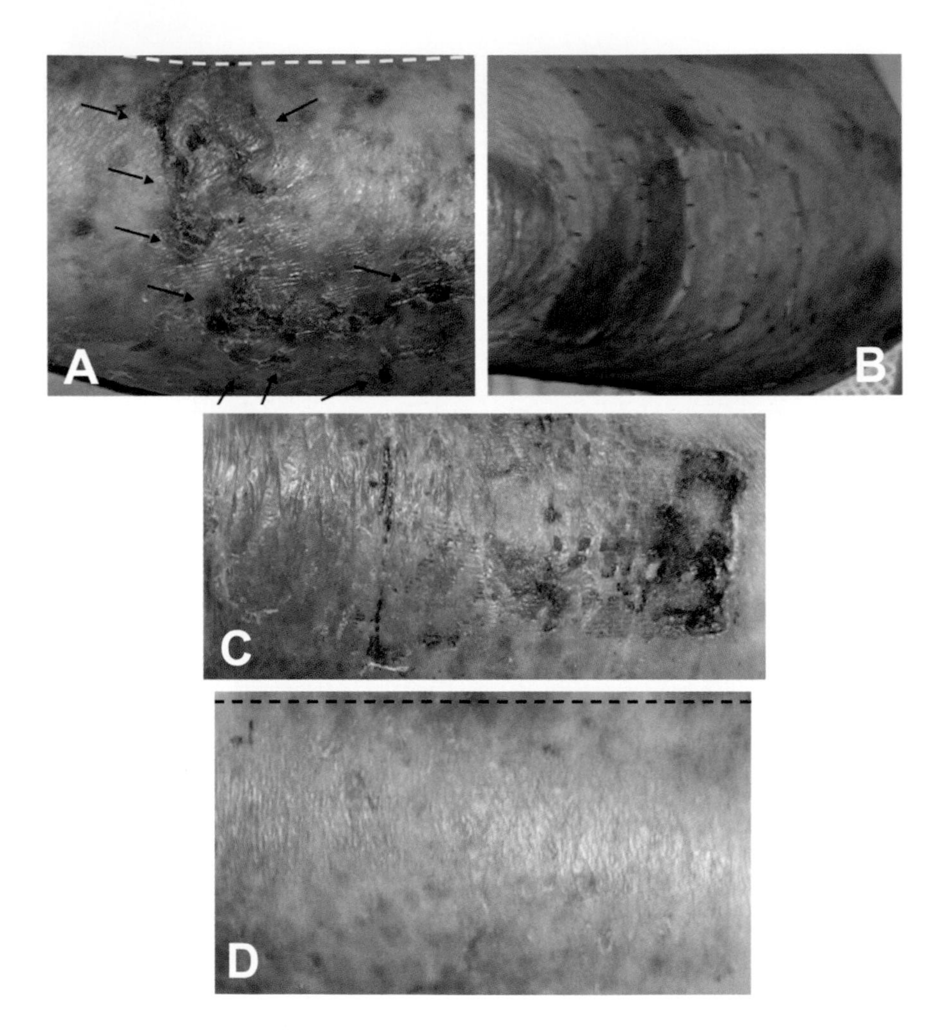


**Figure 27.** Results of gene therapy. (A) Untreated region. Arrows indicate blisters. (B) Excisions followed by application of grafts. (C) Eight days after grafting. (D) Sixty days after grafting. Hatched line in (D) corresponds to hatched line in (A) (Mavilio *et al.*, 2006).

## Chapter Eight

# Therapy with Cultured Chondrocytes

In 1989, Dr. Anders Lindahl was in my laboratory working on the exotic subject of transduced human keratinocytes which, when grafted to athymic mice, secreted growth hormone into the blood of the animals (Lindahl *et al.*, 1990; Teumer *et al.*, 1990).

Upon his return to Sweden, Dr. Lindahl began a collaborative work with Dr. Lars Peterson to study the possibility of treating human traumatic lesions or osteochondritis dissecans of the femoral condyle of the knee with cultured cells. A biopsy of healthy cartilage was obtained from the patient and the tissue was enzymatically dissociated to produce a suspension of single cells, which were expanded in culture by a factor of 10-20 over a period of 2-3 weeks. The lesion in the patient's knee was then excised, a periosteal flap was sutured to the surrounding rim of normal cartilage and the cultured chondrocytes were injected beneath the periosteal flap (Fig. 28) (Brittberg *et al.*, 1994). In a total of 16 treated patients followed for 16-60 months, all were relieved of pain, swelling, crepitation and knee-lock. Two years later, in 14 of the 16 patients the results were graded as either excellent or good. Subsequent biopsies of the regenerated tissue showed normal hyaline cartilage containing type II collagen.

*Therapy with Cultured Cells* by H Green

www.panstanford.com
978-981-4267-70-0

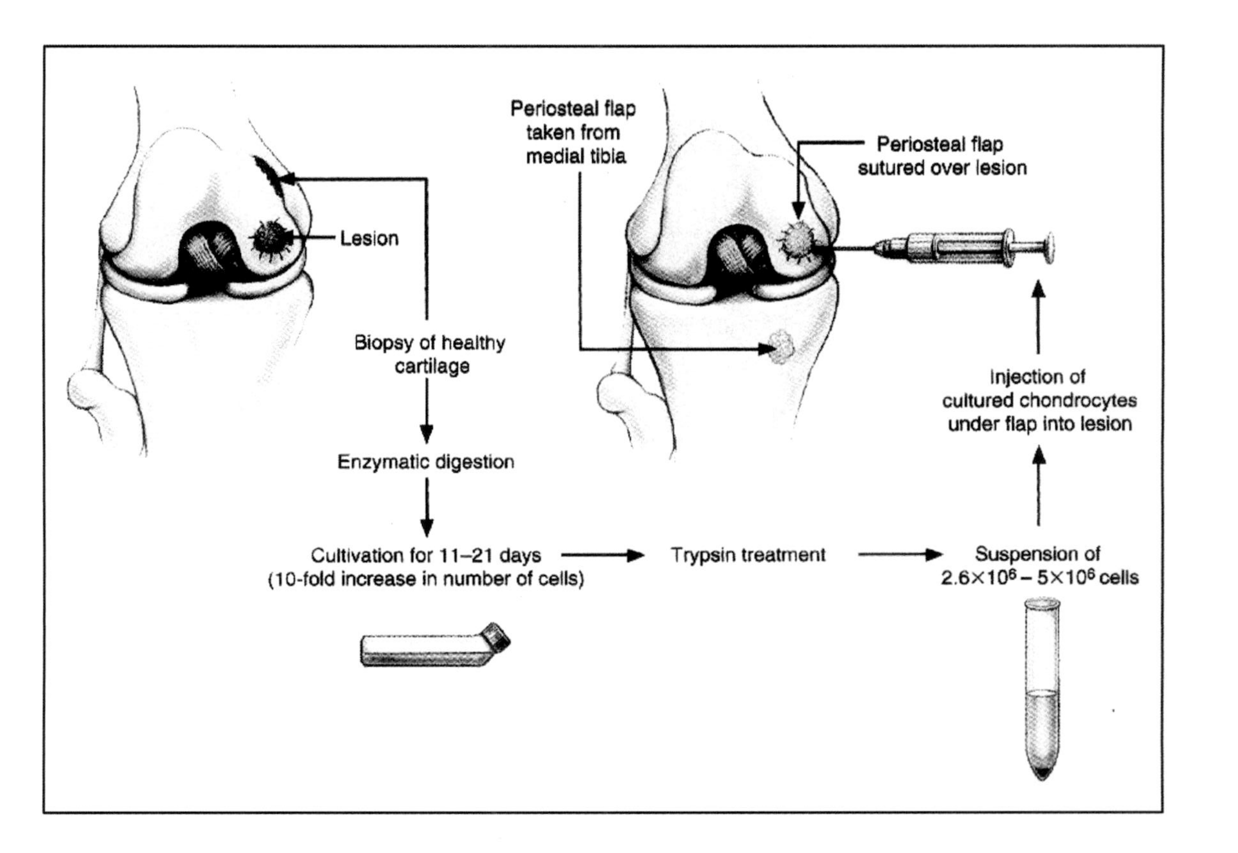


**Figure 28.** Diagram of chondrocyte transplantation in the right femoral condyle. The distal part of the femur and proximal part of the tibia are shown (Brittberg *et al.*, 1994).

In the ensuing 15 years, autologous chondrocyte transplantation has resulted in excellent long-term clinical results for the treatment of cartilage injuries and osteochondral lesions in the knee and ankle (Peterson *et al.*, 2000; Peterson *et al.*, 2002; Petersen *et al.*, 2003; Peterson *et al.*, 2003).

Over the last 15 years, the Göteborg group and other groups in Europe have treated a total of 6700 patients.

Work at Genzyme and by academics in the US and Europe has been in essential agreement with the work of the Göteborg group (Minas 1998; McPherson and Tubo 2000; Micheli *et al.*, 2001; Minas 2001; Bentley *et al.*, 2003; Browne *et al.*, 2005; Fu *et al.*, 2005; Ronga *et al.*, 2005; Micheli *et al.*, 2006; Levine, 2007). Genzyme has treated over 21,000 patients with cultured autologous chondrocytes for the treatment of cartilage injuries and osteochondral lesions.

The use of cultured autologous chondrocytes for different deformities of the nose and ears has been developed in the Yanaga Clinic and Tissue Culture Laboratory, located in Fukuoka, Japan. A total of 92 patients have been treated, with good results (Yanaga *et al.*, 2004; Yanaga *et al.*, 2006). No significant complications were observed.

One example shown in Figure 29 is a case of Microtia, a congenital ear hypoplasia (Yanaga *et al.*, 2009). A sample of cartilage from the ear was grown in culture by methods described earlier (Yanaga *et al.*, 2004; Yanaga *et al.*, 2006). The cartilage was then injection-implanted to the lower abdomen of the patient, where the cells grew into a large cartilage with neoperichondrium in six months. This cartilage was surgically harvested, sculptured into an ear framework and implanted into the position of the new ear. In 2-5 years of post-operative monitoring, the neo-cartilage maintained good shape without absorption.

Research on therapy with cultured chondrocytes is very advanced at the Japan Tissue Engineering Company (J-TEC). Clinical studies were completed two years ago in collaboration with Professor Mitsuo Ochi, of Hiroshima University; but

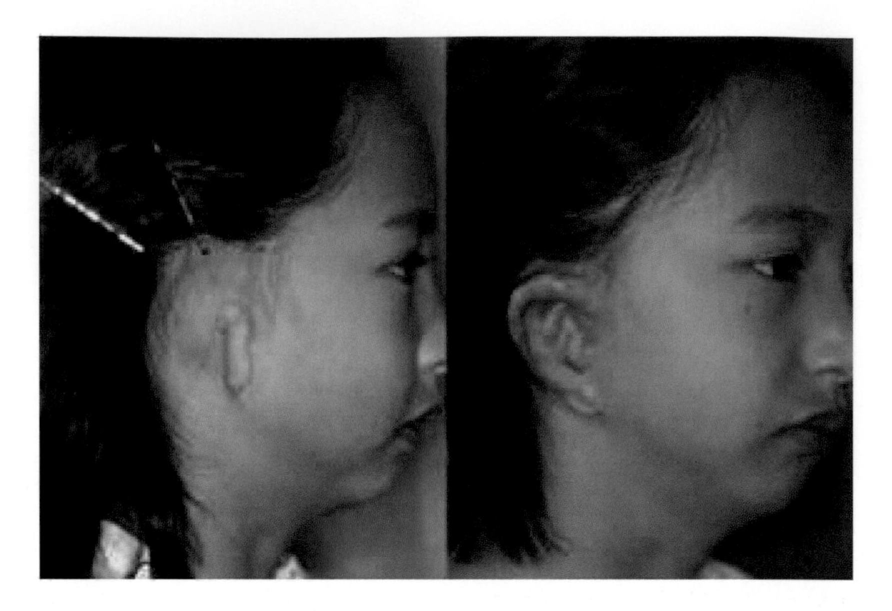

**Figure 29.** A 10-year-old girl. *Left*, Preoperative view. *Right*, four years after implantation to the temporal area and three years after ear elevation. Final appearance and the patient's course were favorable, and there has been no absorption of the cartilage (Yanaga *et al.*, 2009, in press).

the company does not expect to obtain marketing approval for another two years (the regulatory authorities are very slow in Japan).

Chapter Nine

# The Promise of Therapy with Embryonic Stem Cells

Although post-natal keratinocytes, ocular limbal cells and chondrocytes have been successfully used therapeutically, somatic stem cells of other tissues frequently cannot be grown in culture. For this reason many debilitating diseases that could conceivably benefit from cell-based therapy are not currently treatable (Type I diabetes, Parkinson's disease and other neurogenerative disease, cardiomyopathy, etc.).

All somatic cell types can be derived from embryonic stem cells and in principle are available for therapy; but there are two major obstacles that have to be overcome before they can be used therapeutically:

1. The somatic cell types derived from hES cells have limited growth potential.
2. All residual hES cells must be eliminated because they are teratogenic.

The prevalent idea that somatic cell types obtained from the differentiation of hES cells and proposed for human therapy are identical to the corresponding post-natal somatic cells formed by the normal process of embryogenesis is not well founded. Differentiation of hES cells in culture takes place in the absence of developmental cues such as are afforded

*Therapy with Cultured Cells* by H Green

www.panstanford.com
978-981-4267-70-0

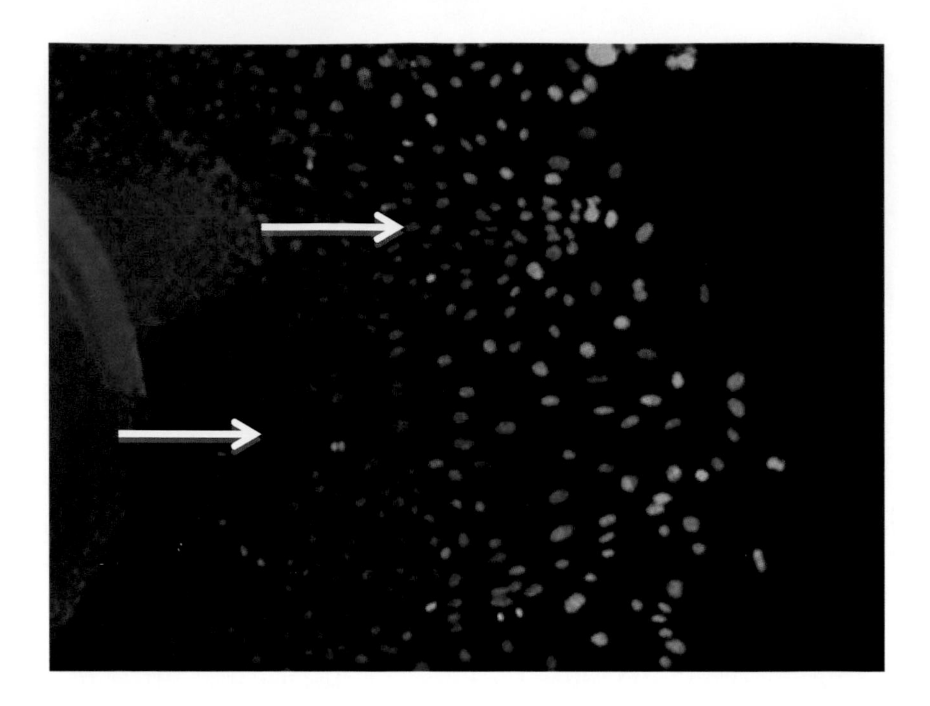

**Figure 30.** A single embryoid body was deposited on a culture dish. All cells in the embryoid body contain Oct4 (red). All cells migrating to the right are shown by nuclear stain, but lack Oct4.

by polarity and gradients and the differentiation takes place much more rapidly than that which occurs during embryogenesis.

## MARKER SUCCESSION DURING THE DEVELOPMENT OF KERATINOCYTES FROM CULTURED HUMAN EMBRYONIC STEM CELLS

For analysis of the development of keratinocytes from human embryonic stem cells it is useful to study the appearance of markers in cells migrating from cultured embryoid bodies (Green *et al.*, 2003). The order of marker succession is as follows:

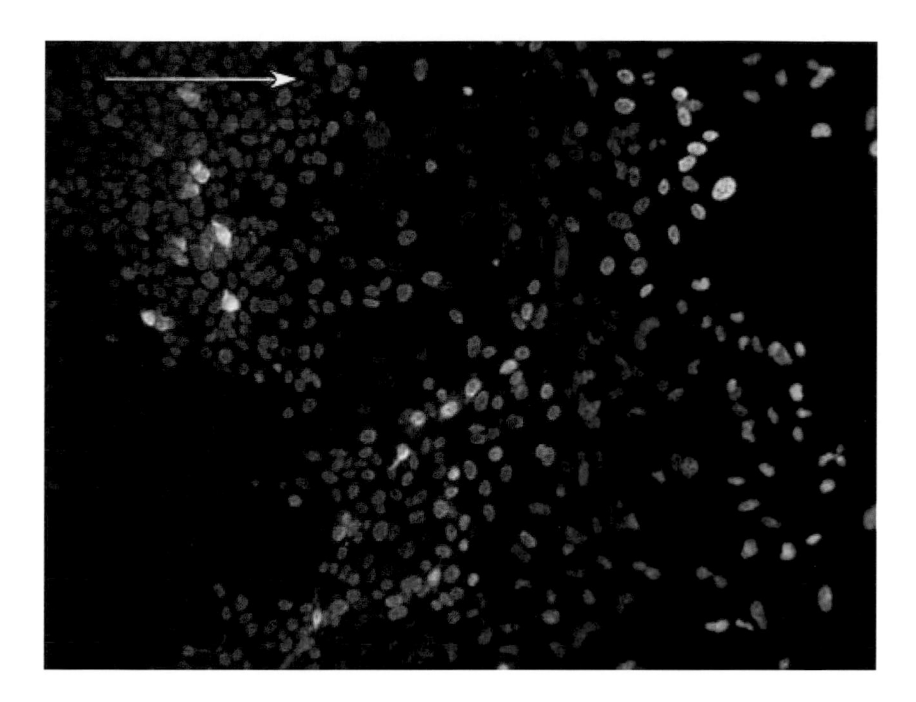

**Figure 31.** Appearance of p63 and K14 from an attached embryoid body inoculated 15 days earlier. Red – nuclear p63; green – cytoplasmic K14; blue – nuclei. No K14 containing cells lack p63.

1. Loss of Oct4. This loss is seen in cells migrating away from an embryoid body (Fig. 30).

2. Appearance of keratinocytes containing p63, K14 and basonuclin, markers of the basal layer of the keratinocytes. This is seen in Figure 31.

3. Multiplication of the keratinocytes and their concentration close to the migration front. This is seen in Figure 32. The presence of basonuclin is known to be associated with proliferative capacity of the keratinocyte (Tseng and Green, 1994).

4. Terminal differentiation. After many days, there appeared scattered squame-like structures containing involucrin, a precursor of the cross-linked envelope characteristic of terminal differentiation (Fig. 33).

One example of the differences between somatic cells derived from hES cells and those produced by the normal process of embryogenesis is seen in the keratinocyte. Post-natal keratinocytes are probably the best growing of all human cell types: they are capable of growing through up to 150 cell generations in culture. hES cell-derived keratinocytes are strikingly different in that they have very little capacity for proliferation.

## ORIGIN OF NOD CELLS

There are two ways of obtaining keratinocytes from hES cells. One is to form embryoid bodies *in vitro*, allow them to undergo differentiation and harvest the resulting keratinocytes. These cells grow very poorly, although they can be made to grow by introducing the E6E7 genes of HPV16 (see below and Iuchi *et al.*, 2006). A more convenient method is to inject hES cells into *scid* mice and allow them to grow into nodules in which differentiation into keratinocytes will take place. We call keratinocytes recovered *in vitro* from such nodules Nod cells to indicate that they are recovered from nodules.

## DIFFERENCES BETWEEN POST-NATAL KERATINOCYTES AND hES CELL-DERIVED KERATINOCYTES

Cultured post-natal human keratinocytes have great proliferative potential in culture. They make coherent colonies and grow rapidly, while excavating the surrounding supporting 3T3 cells (Fig. 34).

The behavior of hES cell-derived keratinocytes is very different (Fig. 35). Their colonies are not coherent: instead they undergo fragmentation into two or more subclones.

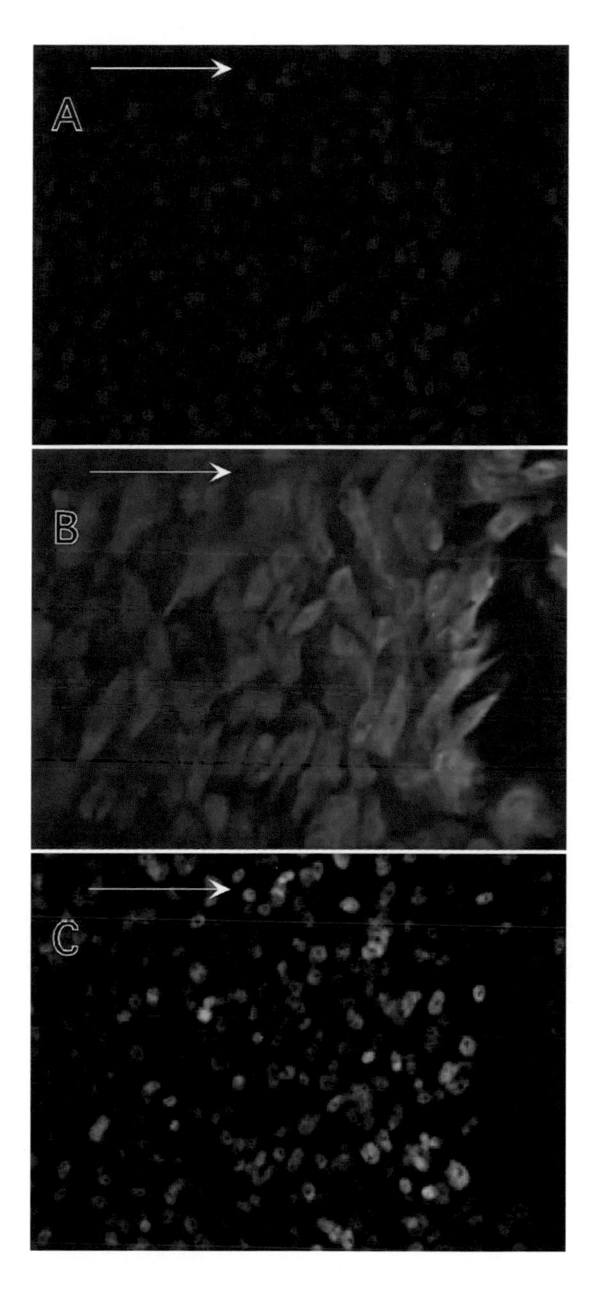


**Figure 32.** After 27 days, the zone close to the migration front is completely composed of cells containing p63 (red), K14 (green) and basonuclin (blue).

**Figure 33.** Appearance of involucrin in a stratified keratinocyte colony in the region of migration from a cultured embryoid body. Nearly all cells contain p63 (red). Most cells contain K14 (green). Scattered squame-like structures overlying the basal layer contain involucrin (blue) (Green *et al.*, 2003).

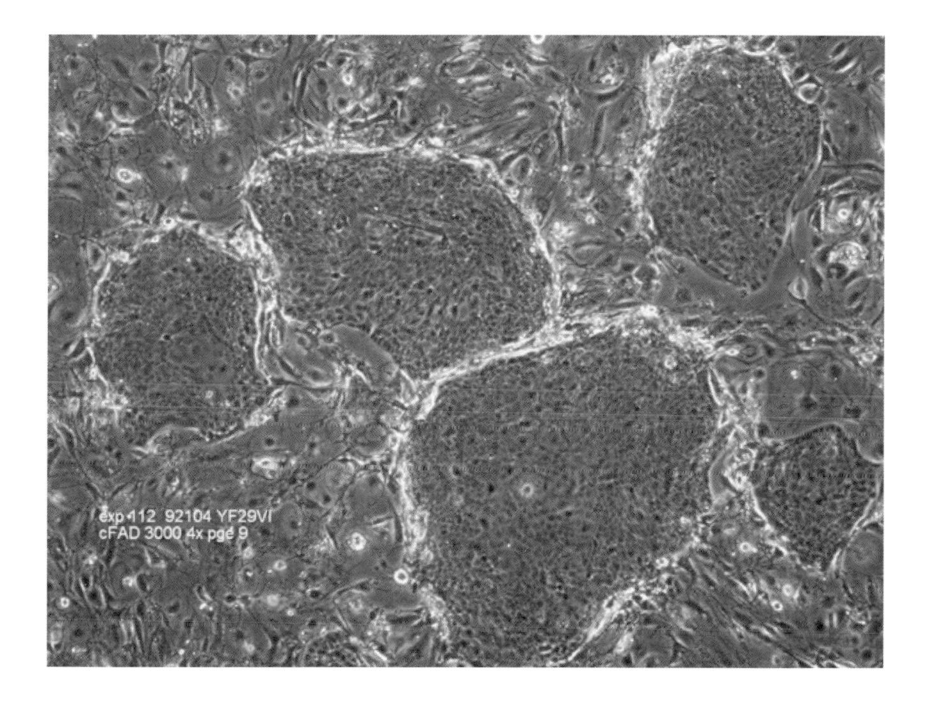


**Figure 34.** Post-natal human keratinocytes. The colonies are composed of coherent keratinocytes, which excavate the surrounding 3T3 cells as the colonies expand.

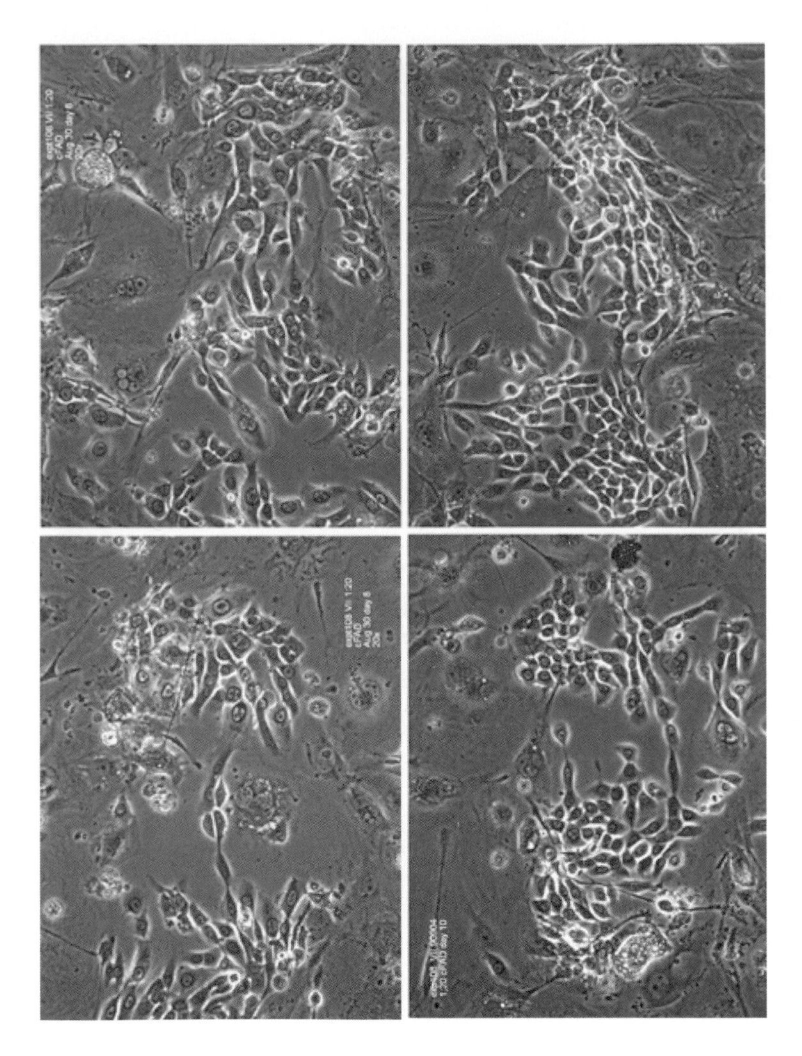

**Figure 35.** Four examples of hES cell-derived keratinocyte colonies undergoing fragmentation.

This is shown even more clearly by staining the colonies for p63 (Fig. 36). The cleavage furrows of the colonies of hES cell-derived keratinocytes are evident and anticipate fragmentation. This behavior is never seen in post-natal keratinocytes and must relate somehow to the fact that the hES cell-derived keratinocytes have such a short proliferative lifetime. Notice that the number of post-natal keratinocytes of colony A at day 7 is much greater than the number of Nod3 cells in colony B at day 12.

## EFFECTS OF TRANSDUCTION WITH THE E6E7 GENES OF HPV16

Expression of the E6E7 genes of HPV16 has been shown to disable p53 and RB-mediated checkpoints (Dyson *et al.* 1989; Scheffner *et al.* 1990; Werness *et al.* 1990; Boyer *et al.* 1996; Jones and Munger 1997) and to induce expression of the endogenous TERT gene in keratinocytes (Klingelhutz *et al.* 1996), thereby providing an efficient method for producing immortalized human keratinocyte lines (Hawley-Nelson *et al.* 1989; Munger *et al.* 1989). We therefore transduced Nod3-derived cells with the L(HPV16E6E7)SN vector and serially passaged the cells without drug selection. The rapidity of action of E6E7 was evident as early as eight days after transduction, when many colonies in the transduced cultures were nearly as large as those of post-natal keratinocytes (Fig. 37). The cells were coherent and could even be used to make a cultured graft (Fig. 38). On day 10, when a culture of each was trypsinized, the transduced culture yielded 6.8 times more keratinocytes than the control. The Nod3/E6E7 cells then grew through over 60 population doublings more than the untransduced control Nod3 cells. As there was no sign of a diminishing growth rate, we concluded that the cells had become immortalized.

E6E7 transduction was also carried out on cells taken from the region of cell migration from an attached embryoid

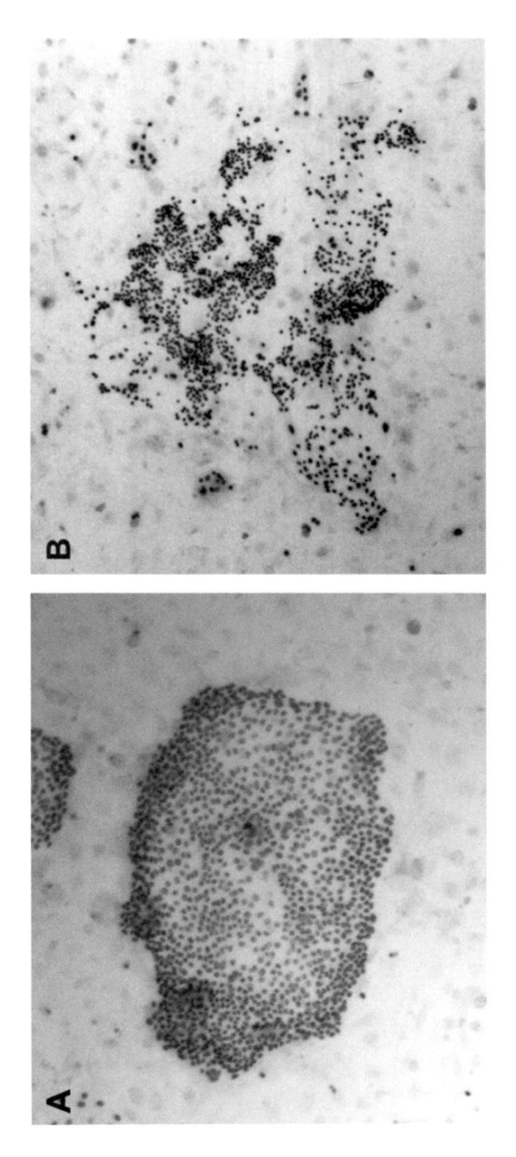


**Figure 36.** (A) 7 day colony of post-natal keratinocytes. (B) 12 day colony of hES cell-derived keratinocytes undergoing fragmentation. Both are stained with antibody to p63.

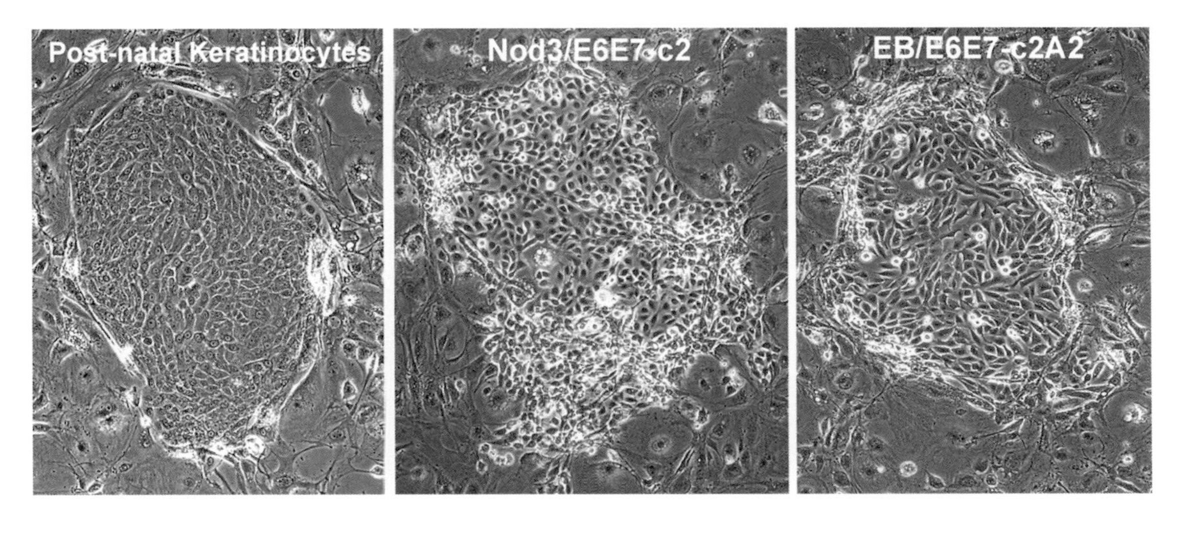


**Figure 37.** Colonies of post-natal and of hES-derived, E6E7 transduced keratinocytes. Eight-day colonies are shown for each cell type using a 10× phase objective. The colony of post-natal keratinocytes (left) is composed of tightly packed cells, with a smooth perimeter at the site of excavation of the 3T3 supporting cells. Colonies of the E6E7 infected clones derived from nodules (center) and from cultured embryoid bodies (right) are coherent, though not as regular as the colony of post-natal keratinocytes.

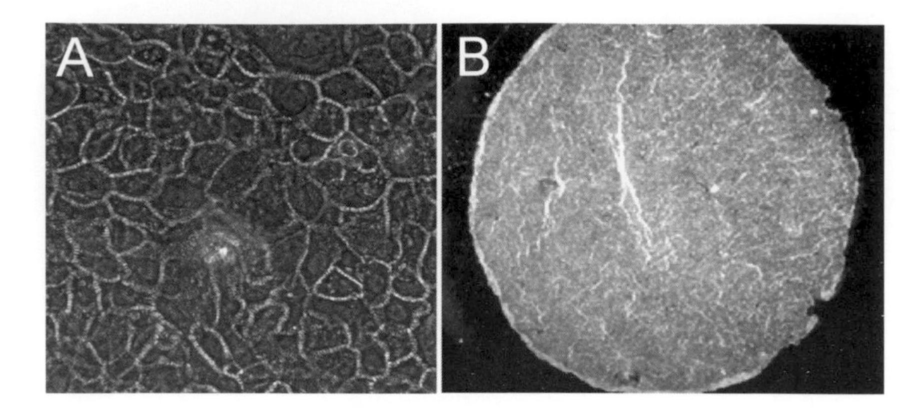


**Figure 38.** Cohesion of Nod3/E6E7-c2 cells. (A) Twelve-day confluent culture, 40× objective, showing — intercellular bridges, an old term used to describe what is seen by light microscopy as a result of the presence of desmosomes. (B) Detachment of the cell layer as an integral sheet by dispase (Iuchi *et al.*, 2006).

body (EB), a region rich in cells bearing keratinocyte markers (Green *et al.*, 2003). Prior to the transduction, such keratinocytes had practically no ability to multiply on subculture. Within days of the transduction, many growing colonies appeared and about one-half had keratinocyte-like morphology (Iuchi *et al.* 2006).

These experiments showed that it is possible to correct the restricted growth potential of hES cell-derived keratinocytes by introduction of exogenous genes. But obviously the genes of an oncogenic virus could not be used to prepare cells for therapeutic use.

We investigated the possibility of improving the multiplication of the hES cell-derived keratinocytes by the introduction of exogenous genes known to promote multiplication, such as CDK4, E2F, BMI-1 and CBX7. Some of these had no effect but others had some effects, even if they were too small to be useful. In principle, these effects demonstrated that the introduction of exogenous genes could result in increased proliferative potential. Studies in the laboratory of James Rheinwald showed that a variety of other genes were able to increase life-span (Dabelsteen *et al.*, 2009). The future of practical applications of hES cell-derived somatic cells depends on the increase of their proliferative potential by the introduction of exogenous genes.

# A Final Philosophical Reflection

There is great enthusiasm among scientists for the practical possibilities of embryonic stem cells in human therapy. This enthusiasm has been communicated to the general public. But I have described the difficulties and the risks involved. It is by no means clear how these problems will be resolved. A policy of prudence and caution would be the best policy.

*Therapy with Cultured Cells* by H Green

www.panstanford.com
978-981-4267-70-0

# References

[1] Albrecht-Buehler, G. (1977). "Phagokinetic tracks of 3T3 cells: parallels between the orientation of track segments and of cellular structures which contain actin or tubulin." *Cell* **12**(2): 333-9.

[2] Allen-Hoffmann, B. L. and J. G. Rheinwald (1984). "Polycyclic aromatic hydrocarbon mutagenesis of human epidermal keratinocytes in culture." *Proceedings of the National Academy of Sciences of the United States of America* **81**(24): 7802-6.

[3] Alvarez-Diaz, C., J. Cuenca-Pardo, A. Sosa-Serrano, E. Juarez-Aguilar, M. Marsch-Moreno and W. Kuri-Harcuch (2000). "Controlled clinical study of deep partial-thickness burns treated with frozen cultured human allogeneic epidermal sheets." *J Burn Care Rehabil* **21**(4): 291-9.

[4] Banks-Schlegel, S. and H. Green (1980). "Formation of epidermis by serially cultivated human epidermal cells transplanted as an epithelium to athymic mice." *Transplantation* **29**(4): 308-13.

[5] Beele, H., J. M. Naeyaert, M. Goeteyn, M. De Mil and A. Kint (1991). "Repeated cultured epidermal allografts in the treatment of chronic leg ulcers of various origins." *Dermatologica* **183**(1): 31-5.

[6] Bentley, G., L. C. Biant, R. W. Carrington, M. Akmal, A. Goldberg, A. M. Williams, J. A. Skinner and J. Pringle (2003). "A prospective, randomised comparison of autologous chondrocyte implantation versus mosaicplasty for osteochondral defects in the knee." *J Bone Joint Surg Br* **85**(2): 223-30.

[7] Blanpain, C., W. E. Lowry, A. Geoghegan, L. Polak and E. Fuchs (2004). "Self-renewal, multipotency, and the existence of two cell populations within an epithelial stem cell niche." *Cell* **118**(5): 635-48.

[8] Bolivar-Flores, Y. J. and W. Kuri-Harcuch (1999). "Frozen allogeneic human epidermal cultured sheets for the cure of complicated leg ulcers." *Dermatol Surg* **25**(8): 610-7.

[9] Boyer, S. N., D. E. Wazer and V. Band (1996). "E7 protein of human papilloma virus-16 induces degradation of retinoblastoma protein through the ubiquitin-proteasome pathway." *Cancer Res* **56**(20): 4620-4.

[10] Brain, A., P. Purkis, P. Coates, M. Hackett, H. Navsaria and I. Leigh (1989). "Survival of cultured allogeneic keratinocytes transplanted to deep dermal bed assessed with probe specific for Y chromosome." *BMJ* **298**(6678): 917-9.

[11] Brittberg, M., A. Lindahl, A. Nilsson, C. Ohlsson, O. Isaksson and L. Peterson (1994). "Treatment of deep cartilage defects in the knee with autologous chondrocyte transplantation." *N Engl J Med* **331**(14): 889-95.

[12] Browne, J. E., A. F. Anderson, R. Arciero, B. Mandelbaum, J. B. Moseley, Jr., L. J. Micheli, F. Fu and C. Erggelet (2005). "Clinical outcome of autologous chondrocyte implantation at 5 years in US subjects." *Clin Orthop Relat Res* (436): 237-45.

[13] Burt, A. M., C. D. Pallett, J. P. Sloane, M. J. O'Hare, K. F. Schafler, P. Yardeni, A. Eldad, J. A. Clarke and B. A. Gusterson (1989). "Survival of cultured allografts in patients with burns assessed with probe specific for Y chromosome." *BMJ* **298**(6678): 915-7.

[14] Carpenter, G. and S. Cohen (1990). "Epidermal growth factor." *J Biol Chem* **265**(14): 7709-12.

[15] Carrel, A. and M. T. Burrows (1910a). "Cultivation of adult tissues and organs outside the body." *J Am Med Assoc* **55**: 1379-1381.

[16] Carrel, A. and M. T. Burrows (1910b). "Culture des tissus adultes en dehors de l'organisme."*Compt Rend Soc Biol* **69**: 293, 298, 299.

[17] Carsin, H., P. Ainaud, H. Le Bever, J. Rives, A. Lakhel, J. Stephanazzi, F. Lambert and J. Perrot (2000). "Cultured epithelial autografts in extensive burn coverage of severely traumatized patients: a five year single-center experience with 30 patients." *Burns* **26**(4): 379-87.

[18] Celli, J., P. Duijf, B. C. Hamel, M. Bamshad, B. Kramer, A. P. Smits, R. Newbury-Ecob, R. C. Hennekam, G. Van Buggenhout, A. van Haeringen, C. G. Woods, A. J. van Essen, R. de

Waal, G. Vriend, D. A. Haber, A. Yang, F. McKeon, H. G. Brunner and H. van Bokhoven (1999). "Heterozygous germline mutations in the p53 homolog p63 are the cause of EEC syndrome." *Cell* **99**(2): 143-53.

[19] Cohen, S. and G. A. Elliott (1963). "The stimulation of epidermal keratinization by a protein isolated from the submaxillary gland of the mouse." *J Invest Dermatol* **40**: 1-5.

[20] Compton, C. C., J. M. Gill, D. A. Bradford, S. Regauer, G. G. Gallico and N. E. O'Connor (1989). "Skin regenerated from cultured epithelial autografts on full-thickness burn wounds from 6 days to 5 years after grafting. A light, electron microscopic and immunohistochemical study." *Lab Invest* **60**(5): 600-12.

[21] Dabelsteen, S., P. Hercule, P. Barron, M. Rice, G. Dorsainville and J. G. Rheinwald (2009). "Epithelial Cells Derived from Human Embryonic Stem Cells Display P16(INK4A) Senescence, Hypermotility, and Differentiation Properties Shared by Many P63(+) Somatic Cell Types." *Stem Cells* **27**(6): 1388-1399.

[22] De Luca, M., E. Albanese, R. Cancedda, A. Viacava, A. Faggioni, G. Zambruno and A. Giannetti (1992). "Treatment of leg ulcers with cryopreserved allogeneic cultured epithelium. A multicenter study." *Arch Dermatol* **128**(5): 633-8.

[23] De Luca, M., G. Pellegrini and H. Green (2006). "Regeneration of squamous epithelia from stem cells of cultured grafts." *Regen Med* **1**(1): 45-57.

[24] Di Iorio, E., V. Barbaro, A. Ruzza, D. Ponzin, G. Pellegrini and M. De Luca (2005). "Isoforms of DeltaNp63 and the migration of ocular limbal cells in human corneal regeneration." *Proc Natl Acad Sci USA* **102**(27): 9523-8.

[25] Duinslaeger, L. A., G. Verbeken, S. Vanhalle and A. Vanderkelen (1997). "Cultured allogeneic keratinocyte sheets accelerate healing compared to Op-site treatment of donor sites in burns." *J Burn Care Rehabil* **18**(6): 545-51.

[26] Dyson, N., P. M. Howley, K. Munger and E. Harlow (1989). "The human papilloma virus-16 E7 oncoprotein is able to bind to the retinoblastoma gene product." *Science* **243**(4893): 934-7.

[27] Eagle, H. (1955). "Nutrition needs of mammalian cells in tissue culture." *Science* **122**(3168): 501-14.

[28] Enders, J. F., T. H. Weller and F. C. Robbins (1949). "Cultivation of the Lansing Strain of Poliomyelitis Virus in Cultures of Various Human Embryonic Tissues." *Science* **109**(2822): 85-87.

[29] Freeman, A. E., P. H. Black, R. Wolford and R. J. Huebner (1967). "Adenovirus type 12-rat embryo transformation system." *J Virol* **1**(2): 362-7.

[30] Fu, F. H., D. Zurakowski, J. E. Browne, B. Mandelbaum, C. Erggelet, J. B. Moseley, Jr., A. F. Anderson and L. J. Micheli (2005). "Autologous chondrocyte implantation versus debridement for treatment of full-thickness chondral defects of the knee: an observational cohort study with 3-year follow-up." *Am J Sports Med* **33**(11): 1658-66.

[31] Gallico, G. G., N. E. O'Connor, C. C. Compton, O. Kehinde and H. Green (1984). "Permanent coverage of large burn wounds with autologous cultured human epithelium." *N Engl J Med* **311**(7): 448-51.

[32] Gallico, G. G., N. E. O'Connor, C. C. Compton, J. P. Remensynder, O. Kehinde and H. Green (1989). "Cultured epithelial autografts for giant congenital nevi." *Plastic & Reconstructive Surgery* **84**: 1-9.

[33] Green, H. (1978). "Cyclic AMP in relation to proliferation of the epidermal cell: a new view." *Cell* **15**(3): 801-11.

[34] Green, H. (1989). "Regeneration of the skin after grafting of epidermal cultures [editorial]." *Lab Invest* **60**(5): 583-4.

[35] Green, H. (2008). "The birth of therapy with cultured cells." *Bioessays* **30**(9): 897-903.

[36] Green, H., K. Easley and S. Iuchi (2003). "Marker succession during the development of keratinocytes from cultured human embryonic stem cells." *Proc Natl Acad Sci USA* **100**(26): 15625-30.

[37] Green, H. and O. Kehinde (1974). "Sublines of mouse 3T3 cells that accumulate lipid." *Cell* **1**: 113-6.

[38] Green, H. and O. Kehinde (1976). "Spontaneous heritable changes leading to increased adipose conversion in 3T3 cells." *Cell* **7**(1): 105-13.

[39] Green, H., O. Kehinde and J. Thomas (1979). "Growth of cultured human epidermal cells into multiple epithelia suitable for grafting." *Proc Natl Acad Sci USA* **76**(11): 5665-8.

[40] Green, H. and M. Meuth (1974). "An established pre-adipose cell line and its differentiation in culture." *Cell* **3**(2): 127-33.

[41] Guerra, L., S. Capurro, F. Melchi, G. Primavera, S. Bondanza, R. Cancedda, A. Luci, M. De Luca and G. Pellegrini (2000). "Treatment of "stable" vitiligo by Timedsurgery and transplantation of cultured epidermal autografts." *Arch Dermatol* **136**(11): 1380-9.

[42] Guerra, L., G. Primavera, D. Raskovic, G. Pellegrini, O. Golisano, S. Bondanza, S. Kuhn, P. Piazza, A. Luci, F. Atzori and M. De Luca (2004). "Permanent repigmentation of piebaldism by erbium:YAG laser and autologous cultured epidermis." *Br J Dermatol* **150**(4): 715-21.

[43] Guerra, L., G. Primavera, D. Raskovic, G. Pellegrini, O. Golisano, S. Bondanza, P. Paterna, G. Sonego, T. Gobello, F. Atzori, P. Piazza, A. Luci and M. De Luca (2003). "Erbium:YAG laser and cultured epidermis in the surgical therapy of stable vitiligo." *Arch Dermatol* **139**(10): 1303-10.

[44] Ham, R. G. (1965). "Clonal Growth of Mammalian Cells in a Chemically Defined, Synthetic Medium." *Proc Natl Acad Sci USA* **53**: 288-93.

[45] Hamilton, W. G. and R. G. Ham (1977). "Clonal growth of chinese hamster cell lines in protein-free media." *In Vitro* **13**(9): 537-47.

[46] Harrison, R. G. (1907). "Observations on the living developing nerve fiber." *Proc Soc Exper Biol and Med* **4**: 140-143.

[47] Harvima, I. T., S. Virnes, L. Kauppinen, M. Huttunen, P. Kivinen, L. Niskanen and M. Horsmanheimo (1999). "Cultured allogeneic skin cells are effective in the treatment of chronic diabetic leg and foot ulcers." *Acta Derm Venereol* **79**(3): 217-20.

[48] Hawley-Nelson, P., K. H. Vousden, N. L. Hubbert, D. R. Lowy and J. T. Schiller (1989). "HPV16 E6 and E7 proteins cooperate

to immortalize human foreskin keratinocytes." *EMBO Journal* **8**(12): 3905-10.

[49] Hayashi, I. and G. H. Sato (1976). "Replacement of serum by hormones permits growth of cells in a defined medium." *Nature* **259**(5539): 132-4.

[50] Hayflick, L. and P. S. Moorhead (1961). "The serial cultivation of human diploid cell strains." *Exp Cell Res* **25**: 585-621.

[51] Hefton, J. M., D. Caldwell, D. G. Biozes, A. K. Balin and D. M. Carter (1986). "Grafting of skin ulcers with cultured autologous epidermal cells." *J Am Acad Dermatol* **14**(3): 399-405.

[52] Hefton, J. M., M. R. Madden, J. L. Finkelstein and G. T. Shires (1983). "Grafting of burn patients with allografts of cultured epidermal cells." *Lancet* **2**(8347): 428-30.

[53] Iuchi, S., S. Dabelsteen, K. Easley, J. G. Rheinwald and H. Green (2006). "Immortalized keratinocyte lines derived from human embryonic stem cells." *Proc Natl Acad Sci USA* **103**(6): 1792-7.

[54] Jeon, S., P. Djian and H. Green (1998). "Inability of keratinocytes lacking their specific transglutaminase to form cross-linked envelopes: absence of envelopes as a simple diagnostic test for lamellar ichthyosis." *Proc Natl Acad Sci USA* **95**(2): 687-90.

[55] Jones, D. L. and K. Munger (1997). "Analysis of the p53-mediated G1 growth arrest pathway in cells expressing the human papillomavirus type 16 E7 oncoprotein." *J Virol* **71**(4): 2905-12.

[56] Kenyon, K. R. and S. C. Tseng (1989). "Limbal autograft transplantation for ocular surface disorders." *Ophthalmology* **96**(5): 709-22; discussion 722-3.

[57] Khachemoune, A., Y. M. Bello and T. J. Phillips (2002). "Factors that influence healing in chronic venous ulcers treated with cryopreserved human epidermal cultures." *Dermatol Surg* **28**(3): 274-80.

[58] Klingelhutz, A. J., S. A. Foster and J. K. McDougall (1996). "Telomerase activation by the E6 gene product of human papillomavirus type 16." *Nature* **380**(6569): 79-82.

[59] Kohler, G. and C. Milstein (1975). "Continuous cultures of fused cells secreting antibody of predefined specificity." *Nature* **256**(5517): 495-7.

[60] Kumagai, N., H. Nishina, H. Tanabe, T. Hosaka, H. Ishida and Y. Ogino (1988). "Clinical application of autologous cultured epithelia for the treatment of burn wounds and burn scars." *Plast Reconstr Surg* **82**(1): 99-110.

[61] Leigh, I. M., P. E. Purkis, H. A. Navsaria and T. J. Phillips (1987). "Treatment of chronic venous ulcers with sheets of cultured allogenic keratinocytes." *Br J Dermatol* **117**(5): 591-7.

[62] Levine, D. W. (2007). Tissue-Engineered Cartilage Products. *Principles of Tissue Engineering, 3rd Edition*. R. P. Lanza, R. Langer and J. Vacanti. Burlington, MA, Elsevier, Inc.: 1215-1224.

[63] Lindahl, A., J. Teumer and H. Green (1990). Cellular aspects of gene therapy. *Growth Factors in Health and Disease*. B. Westermark, C. Betsholtz and B. Hökfelt, Elsevier Science Publishers, B.V.: 383-92.

[64] Lindberg, K., M. E. Brown, H. V. Chaves, K. R. Kenyon and J. G. Rheinwald (1993). "In vitro propagation of human ocular surface epithelial cells for transplantation." *Invest Ophthalmol Vis Sci* **34**(9): 2672-9.

[65] Macpherson, I. and L. Montagnier (1964). "Agar Suspension Culture for the Selective Assay of Cells Transformed by Polyoma Virus." *Virology* **23**: 291-4.

[66] Marcusson, J. A., C. Lindgren, A. Berghard and R. Toftgard (1992). "Allogeneic cultured keratinocytes in the treatment of leg ulcers. A pilot study." *Acta Derm Venereol* **72**(1): 61-4.

[67] Mavilio, F., G. Pellegrini, S. Ferrari, F. Di Nunzio, E. Di Iorio, A. Recchia, G. Maruggi, G. Ferrari, E. Provasi, C. Bonini, S. Capurro, A. Conti, C. Magnoni, A. Giannetti and M. De Luca (2006). "Correction of junctional epidermolysis bullosa by transplantation of genetically modified epidermal stem cells." *Nat Med* **12**(12): 1397-402.

[68] McKeon, F. (2004). "p63 and the epithelial stem cell: more than status quo?" *Genes Dev* **18**(5): 465-9.

[69] McPherson, J. M. and R. Tubo (2000). Articular Cartilage Injury. *Principles of Tissue Engineering, 2nd Edition*. R. P. Lanza, R. Langer and J. Vacanti. San Diego, CA, Academic Press: 697-709.

[70] Micheli, L. J., J. E. Browne, C. Erggelet, F. Fu, B. Mandelbaum, J. B. Moseley and D. Zurakowski (2001). "Autologous chondrocyte implantation of the knee: multicenter experience and minimum 3-year follow-up." *Clin J Sport Med* **11**(4): 223-8.

[71] Micheli, L. J., J. B. Moseley, A. F. Anderson, J. E. Browne, C. Erggelet, R. Arciero, F. H. Fu and B. R. Mandelbaum (2006). "Articular cartilage defects of the distal femur in children and adolescents: treatment with autologous chondrocyte implantation." *J Pediatr Orthop* **26**(4): 455-60.

[72] Miller, D. A., O. J. Miller, V. G. Dev, S. Hashmi, R. Tantravahi, L. Medrano and H. Green (1974). "Human chromosome 19 carries a polio virus receptor gene." *Cell* **1**: 167-73.

[73] Mills, A. A., B. Zheng, X. J. Wang, H. Vogel, D. R. Roop and A. Bradley (1999). "p63 is a p53 homologue required for limb and epidermal morphogenesis." *Nature* **398**(6729): 708-13.

[74] Minas, T. (1998). "Chondrocyte implantation in the repair of chondral lesions of the knee: economics and quality of life." *Am J Orthop* **27**(11): 739-44.

[75] Minas, T. (2001). "Autologous chondrocyte implantation for focal chondral defects of the knee." *Clin Orthop Relat Res* (391 Suppl): S349-61.

[76] Munger, K., W. C. Phelps, V. Bubb, P. M. Howley and R. Schlegel (1989). "The E6 and E7 genes of the human papillomavirus type 16 together are necessary and sufficient for transformation of primary human keratinocytes." *J Virol* **63**(10): 4417-21.

[77] O'Connor, N. E., J. B. Mulliken, S. Banks-Schlegel, O. Kehinde and H. Green (1981). "Grafting of burns with cultured epithelium prepared from autologous epidermal cells." *Lancet* **Jan 10; 1**: 75-8.

[78] Parsa, R., A. Yang, F. McKeon and H. Green (1999). "Association of p63 with proliferative potential in normal and neoplastic human keratinocytes." *J Invest Dermatol* **113**(6): 1099-105.

[79] Pellegrini, G., P. Rama, F. Mavilio and M. De Luca (2009). "Epithelial stem cells in corneal regeneration and epidermal gene therapy." *J Pathol* **217**(2): 217-28.

[80] Pellegrini, G., R. Ranno, G. Stracuzzi, S. Bondanza, L. Guerra, G. Zambruno, G. Micali and M. De Luca (1999). "The control of epidermal stem cells (holoclones) in the treatment of massive full-thickness burns with autologous keratinocytes cultured on fibrin." *Transplantation* **68**(6): 868-79.

[81] Pellegrini, G., C. E. Traverso, A. T. Franzi, M. Zingirian, R. Cancedda and M. De Luca (1997). "Long-term restoration of damaged corneal surfaces with autologous cultivated corneal epithelium." *Lancet* **349**(9057): 990-3.

[82] Petersen, L., M. Brittberg and A. Lindahl (2003). "Autologous chondrocyte transplantation of the ankle." *Foot Ankle Clin* **8**(2): 291-303.

[83] Peterson, L., M. Brittberg, I. Kiviranta, E. L. Akerlund and A. Lindahl (2002). "Autologous chondrocyte transplantation. Biomechanics and long-term durability." *Am J Sports Med* **30**(1): 2-12.

[84] Peterson, L., T. Minas, M. Brittberg and A. Lindahl (2003). "Treatment of osteochondritis dissecans of the knee with autologous chondrocyte transplantation: results at two to ten years." *J Bone Joint Surg Am* **85-A Suppl 2**: 17-24.

[85] Peterson, L., T. Minas, M. Brittberg, A. Nilsson, E. Sjogren-Jansson and A. Lindahl (2000). "Two- to 9-year outcome after autologous chondrocyte transplantation of the knee." *Clin Orthop Relat Res* (374): 212-34.

[86] Phillips, T. J., J. Bhawan, I. M. Leigh, H. J. Baum and B. A. Gilchrest (1990). "Cultured epidermal autografts and allografts: a study of differentiation and allograft survival." *J Am Acad Dermatol* **23**(2 Pt 1): 189-98.

[87] Phillips, T. J., M. Bigby and L. Bercovitch (1991). "Cultured allografts as an adjunct to the medical treatment of problematic leg ulcers." *Arch Dermatol* **127**(6): 799-801.

[88] Phillips, T. J. and B. A. Gilchrest (1989). "Cultured allogenic keratinocyte grafts in the management of wound healing: prognostic factors." *J Dermatol Surg Oncol* **15**(11): 1169-76.

[89] Phillips, T. J., O. Kehinde, H. Green and B. A. Gilchrest (1989). "Treatment of skin ulcers with cultured epidermal allografts." *J Am Acad Dermatol* **21**(2 Pt 1): 191-9.

[90] Rama, P., S. Bonini, A. Lambiase, O. Golisano, P. Paterna, M. De Luca and G. Pellegrini (2001). "Autologous fibrin-cultured limbal stem cells permanently restore the corneal surface of patients with total limbal stem cell deficiency." *Transplantation* **72**(9): 1478-85.

[91] Rheinwald, J. G. and H. Green (1975a). "Formation of a keratinizing epithelium in culture by a cloned cell line derived from a teratoma." *Cell* **6**(3): 317-30.

[92] Rheinwald, J. G. and H. Green (1975b). "Serial cultivation of strains of human epidermal keratinocytes: the formation of keratinizing colonies from single cells." *Cell* **6**(3): 331-43.

[93] Rheinwald, J. G. and H. Green (1977). "Epidermal growth factor and the multiplication of cultured human epidermal keratinocytes." *Nature* **265**(5593): 421-4.

[94] Rice, R. H. and H. Green (1977). "The cornified envelope of terminally differentiated human epidermal keratinocytes consists of cross-linked protein." *Cell* **11**(2): 417-22.

[95] Rivas-Torres, M. T., D. Amato, H. Arambula-Alvarez and W. Kuri-Harcuch (1996). "Controlled clinical study of skin donor sites and deep partial-thickness burns treated with cultured epidermal allografts." *Plast Reconstr Surg* **98**(2): 279-87.

[96] Romagnoli, G., M. De Luca, F. Faranda, R. Bandelloni, A. T. Franzi, F. Cataliotti and R. Cancedda (1990). "Treatment of posterior hypospadias by the autologous graft of cultured urethral epithelium." *N Engl J Med* **323**(8): 527-30.

[97] Romagnoli, G., M. De Luca, F. Faranda, A. T. Franzi and R. Cancedda (1993). "One-step treatment of proximal hypospadias by the autologous graft of cultured urethral epithelium." *J Urol* **150**(4): 1204-7.

[98] Ronfard, V., H. Broly, V. Mitchell, J. P. Galizia, D. Hochart, E. Chambon, P. Pellerin and J. J. Huart (1991). "Use of human keratinocytes cultured on fibrin glue in the treatment of burn wounds." *Burns* **17**(3): 181-4.

[99] Ronfard, V., J. M. Rives, Y. Neveux, H. Carsin and Y. Barrandon (2000). "Long-term regeneration of human epidermis on third degree burns transplanted with autologous cultured epithelium grown on a fibrin matrix." *Transplantation* **70**(11): 1588-98.

[100] Ronga, M., F. A. Grassi, C. Montoli, P. Bulgheroni, E. Genovese and P. Cherubino (2005). "Treatment of deep cartilage defects of the ankle with matrix-induced autologous chondrocyte implantation (MACI)." *Foot and Ankle Surgery* **11**(1): 29-33.

[101] Roseeuw, D. I., A. De Coninck, W. Lissens, E. Kets, I. Liebaers, A. Vercruysse and Y. Vandenberghe (1990). "Allogeneic cultured epidermal grafts heal chronic ulcers although they do not remain as proved by DNA analysis." *J Dermatol Sci* **1**(4): 245-52.

[102] Rous, P. and F. S. Jones (1916). "A method for obtaining suspensions of living cells from the fixed tissues, and for the plating out of individual cells." *J. Exp. Med.* **23**(4): 549-555.

[103] Sanford, K. K., W. R. Earle and G. D. Likely (1948). "The growth in vitro of single isolated tissue cells." *J Natl Cancer Inst* **9**(3): 229-46.

[104] Scheffner, M., B. A. Werness, J. M. Huibregtse, A. J. Levine and P. M. Howley (1990). "The E6 oncoprotein encoded by human papillomavirus types 16 and 18 promotes the degradation of p53." *Cell* **63**(6): 1129-36.

[105] Scherer, W. F., J. T. Syverton and G. O. Gey (1953). "Studies on the propagation in vitro of poliomyelitis viruses. IV. Viral multiplication in a stable strain of human malignant epithelial cells (strain HeLa) derived from an epidermoid carcinoma of the cervix." *J Exp Med* **97**(5): 695-710.

[106] Schermer, A., S. Galvin and T. T. Sun (1986). "Differentiation-related expression of a major 64K corneal keratin in vivo and in culture suggests limbal location of corneal epithelial stem cells." *J Cell Biol* **103**(1): 49-62.

[107] Senoo, M., F. Pinto, C. P. Crum and F. McKeon (2007). "p63 Is essential for the proliferative potential of stem cells in stratified epithelia." *Cell* **129**(3): 523-36.

[108] Shehade, S., J. Clancy, A. Blight, K. Young and P. Levick (1989). "Cultured epithelial allografting of leg ulcers." *Journal of Dermatological Treatment* **1**(2): 79 - 81.

[109] Stevens, L. C. (1970). "The development of transplantable teratocarcinomas from intratesticular grafts of pre- and postimplantation mouse embryos." *Dev Biol* **21**(3): 364-82.

[110] Tamariz, E., M. Marsch-Moreno, F. Castro-Munozledo, V. Tsutsumi and W. Kuri-Harcuch (1999). "Frozen cultured sheets of human epidermal keratinocytes enhance healing of full-thickness wounds in mice." *Cell Tissue Res* **296**(3): 575-85.

[111] Teepe, R. G., R. Koch and B. Haeseker (1993a). "Randomized trial comparing cryopreserved cultured epidermal allografts with tulle-gras in the treatment of split-thickness skin graft donor sites." *J Trauma* **35**(6): 850-4.

[112] Teepe, R. G., E. J. Koebrugge, M. Ponec and B. J. Vermeer (1990). "Fresh versus cryopreserved cultured allografts for the treatment of chronic skin ulcers." *Br J Dermatol* **122**(1): 81-9.

[113] Teepe, R. G., D. I. Roseeuw, J. Hermans, E. J. Koebrugge, T. Altena, A. de Coninck, M. Ponec and B. J. Vermeer (1993b). "Randomized trial comparing cryopreserved cultured epidermal allografts with hydrocolloid dressings in healing chronic venous ulcers." *J Am Acad Dermatol* **29**(6): 982-8.

[114] Temin, H. M. and H. Rubin (1958). "Characteristics of an assay for Rous sarcoma virus and Rous sarcoma cells in tissue culture." *Virology* **6**(3): 669-88.

[115] Teumer, J., A. Lindahl and H. Green (1990). "Human growth hormone in the blood of athymic mice grafted with cultures of hormone-secreting human keratinocytes." *Faseb J* **4**(14): 3245-50.

[116] Thivolet, J., M. Faure, A. Demidem and G. Mauduit (1986). "Long-term survival and immunological tolerance of human epidermal allografts produced in culture." *Transplantation* **42**(3): 274-80.

[117] Todaro, G. and H. Green (1963). "Quantitative studies of the growth of mouse embyro cells in culture and their development into established lines." *J Cell Biol* **17**: 299-313.

[118] Todaro, G. J. and H. Green (1964). "An assay for cellular transformation by SV40." *Virology* **23**(1): 117-119.

[119] Tseng, H. and H. Green (1994). "Association of basonuclin with ability of keratinocytes to multiply and with absence of terminal differentiation." *J Cell Biol* **126**(2): 495-506.

[120] van Bokhoven, H. and F. McKeon (2002). "Mutations in the p53 homolog p63: allele-specific developmental syndromes in humans." *Trends Mol Med* **8**(3): 133-9.

[121] Vanhoutteghem, A., P. Djian and H. Green (2008). "Ancient origin of the gene encoding involucrin, a precursor of the cross-linked envelope of epidermis and related epithelia." *Proc Natl Acad Sci USA* **105**(40): 15481-6.

[122] Weiss, M. C. and H. Green (1967). "Human-mouse hybrid cell lines containing partial complements of human chromosomes and functioning human genes." *Proc Natl Acad Sci USA* **58**(3): 1104-11.

[123] Werness, B. A., A. J. Levine and P. M. Howley (1990). "Association of human papillomavirus types 16 and 18 E6 proteins with p53." *Science* **248**(4951): 76-9.

[124] Yanaga, H., I. Keisuke, T. Fujimoto and K. Yanaga (2009). "Generating ears from cultured autologous auricular chondrocytes by using two-stage implantation in microtia treatment." *Plastic & Reconstructive Surgery* September, 2009.

[125] Yanaga, H., M. Koga, K. Imai and K. Yanaga (2004). "Clinical application of biotechnically cultured autologous chondrocytes as novel graft material for nasal augmentation." *Aesthetic Plast Surg* **28**(4): 212-21.

[126] Yanaga, H., Y. Udoh, T. Yamauchi, M. Yamamoto, K. Kiyokawa, Y. Inoue and Y. Tai (2001). "Cryopreserved cultured epidermal allografts achieved early closure of wounds and reduced scar formation in deep partial-thickness burn wounds (DDB) and split-thickness skin donor sites of pediatric patients." *Burns* **27**(7): 689-98.

[127] Yanaga, H., K. Yanaga, K. Imai, M. Koga, C. Soejima and K. Ohmori (2006). "Clinical application of cultured autologous human auricular chondrocytes with autologous serum for craniofacial or nasal augmentation and repair." *Plast Reconstr Surg* **117**(6): 2019-30; discussion 2031-2.

[128] Yang, A., M. Kaghad, Y. Wang, E. Gillett, M. D. Fleming, V. Dotsch, N. C. Andrews, D. Caput and F. McKeon (1998). "p63, a p53 homolog at 3q27-29, encodes multiple products with transactivating, death-inducing, and dominant-negative activities." *Mol Cell* **2**(3): 305-16.

[129] Yang, A., R. Schweitzer, D. Sun, M. Kaghad, N. Walker, R. T. Bronson, C. Tabin, A. Sharpe, D. Caput, C. Crum and F. McKeon (1999). "p63 is essential for regenerative proliferation in limb, craniofacial and epithelial development." *Nature* **398**(6729): 714-8.

---

The author wishes to acknowledge the valuable advice of Dr. Guenter Albrecht–Buehler on the writing of this book.

---

# Index